Wizard Books
Luton Sixth Form College

This edition published in the UK in 2005
by Wizard Books, an imprint of
Icon Books Ltd., The Old Dairy,
Brook Road, Thriplow,
Cambridge SG8 7RG
email: wizard@iconbooks.co.uk
www.iconbooks.co.uk/wizard

Originally published in 2003 by Wizard Books,
an imprint of Icon Books Ltd.

Sold in the UK, Europe, South Africa and Asia
by Faber & Faber Ltd., 3 Queen Square,
London WC1N 3AU
or their agents

Distributed in the UK, Europe, South Africa and Asia
by TBS Ltd., Frating Distribution Centre, Colchester Road
Frating Green, Colchester CO7 7DW

This edition published in Australia in 2005
by Allen & Unwin Pty Ltd.,
PO Box 8500, 83 Alexander Street,
Crows Nest, NSW 2065

Distributed in Canada by
Penguin Books Canada,
90 Eglinton Avenue East, Suite 700,
Toronto, Ontario M4P 2Y3

ISBN 1 84046 661 8

Printed and bound in the UK by Ashford Colour Press Ltd.

One, two, three . . . lots

How big does a number have to be before we call it a Big Number? You might think that a hundred is big, or a thousand, or a million. But among some groups of people, before modern ideas about counting spread around the world from Europe, it was possible to count only up to three. They had no way to describe any larger numbers, except by saying 'lots'. Their counting system went 'one, two, three, lots'. If we counted like that, it would make maths lessons a lot easier, but it would be difficult to cope with the modern world.

What we mean by a Big Number depends on what we are measuring. A thousand people would be a big crowd, but it would be a very small number of atoms, or bacteria. The Big Numbers in this book are all related to what it is they are measuring. Some are bigger than others, but they are all big in their own way.

The scale of space

Space is so big that astronomers have to use special kinds of measurements to describe how far away things are. On Earth, 1,000 kilometres would be a long distance. But a beam of light covers 300,000 kilometres in a single second. The speed of light is always the same, everywhere in space, so it's used to measure distances across the Universe. The Sun is 150 million kilometres away from us, and light covers that distance in 499 seconds. So the distance from the Earth to the Sun can be described as 499 light seconds, or 8.3 light minutes. In a year, light covers 9,460,000,000,000 km. This distance is known as a light year – a light year is a measurement of *distance*, not a measurement of time.

Light years are useful for measuring distances to other stars. The nearest star to the Sun is 4.29 light years away. This looks like a small number when we measure in light years, but it's definitely a big number – 405,834,000,000,000 km – when we measure in everyday units. It's always important to choose the right units when measuring things. It would take a gigantic amount of numbers to measure distances to stars in kilometres, so these vast distances are measured in light years. If we tried to measure distances on Earth in light years, it would mean that instead of saying that the walk to school is 3 km long, we could say that it is 0.00001 light seconds long!

In a year, light covers 9,460,000,000,000 km

5

Our back yard

Light years are even too big to measure distances to planets. The planets orbit around the Sun in a family called the Solar System, which is like our back yard in space. Sometimes, it's useful to think about these distances in terms of light minutes, or light seconds. Radio waves travel at the speed of light. If a space-probe was approaching the planet Jupiter, it would be about 40 light minutes away from us. This would mean that it would take 40 minutes for radio signals from the spacecraft to get back to Earth, and 40 minutes for instructions to the space-probe to get to it from Earth. If the spacecraft controllers here on Earth wanted to instruct the space-probe to fire its rockets at a certain time, they would have to

send the order 40 minutes before the rockets have to fire.

But astronomers usually use another unit to measure distances across the Solar System. They call the distance from the Earth to the Sun one astronomical unit, or 1 AU. So 1 AU is 150 million kilometres. Actually, to be precise it is 149,597,870 km. Distances to the planets are known very accurately because we can bounce radar echoes (which also travel at the speed of light) off Venus and Mars, and measure how long it takes for the echoes to get back. The size of the astronomical unit is calculated from these measurements and the geometry of the Solar System.

Eagles, hawks and falcons are all predatory birds that can see up to eight times more clearly than the human eye. Predator birds live by hunting and killing small animals for food, so having such remarkably sharp sight enables them to see their prey from an incredible distance away with their naked eye. When in flight, a golden eagle can spot a hare from a mile – over a kilometre – away. Hawks' fantastically good eyesight evolved because they needed not only to see their prey from far away, but also to maintain a clear focus while flying at high speeds. Over thousands of generations, predator birds with the best eyesight found the most food and survived, and this meant that they developed better and better eyesight. To survive, find food and breed, birds rely very much on their sense of vision, and if you look at a hawk's eyes you can see how much larger they are in proportion to the rest of its head than your eyes are in your face.

Most birds have monocular vision – by having one eye on either side of their head, they can see things to the side of them as well as in front. Having a wide field of vision like this helps them to see danger as quickly as possible. Pigeons can see through 300 degrees – almost right round their heads – and a bird called a woodcock has its eyes so far back on the side of its head that it can see everything that is going on behind it.

Who can see furthest with the naked eye?

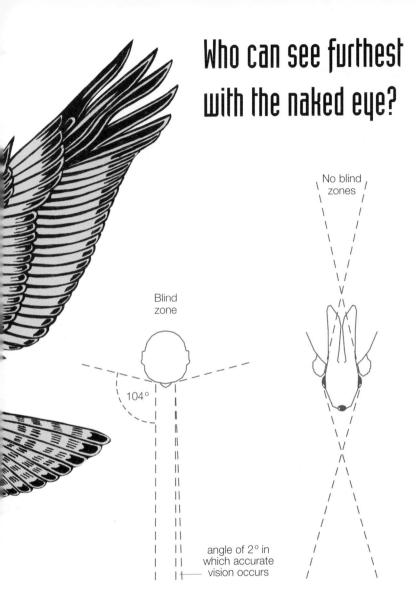

Blind
zone

104°

angle of 2° in
which accurate
vision occurs

No blind
zones

Range of vision of a human

Range of vision of a hare

Rods and cones

Owls have binocular vision. Their eyes are in the front of their flat faces, but they make up for this low field of vision by being able to turn their heads right round to face backwards. The structure of a bird's eye is similar to that of humans, but flatter, and this enables birds to have a larger area in focus at any one time. Nocturnal birds have huge pupils which allow more light to come into their eyes. At the back of every eye there is an area called the retina which is made up of cells called rod cells and cone cells. Rod cells are very sensitive to light, and birds that hunt in the dark have many more of these than humans or birds that hunt during the day. Humans have about 200,000 rod cells in a square millimetre, but an owl has 1,000,000 in the same space, which is why it can see so much better than us at night. Cone cells enable brains to perceive colours. Nocturnal birds don't have many cone cells because they need to see in very low light but don't need to be able to distinguish colours. Humans need to see colours in daylight, and the human eye has about 10,000 cones per square millimetre. Humming-birds that are among the smallest birds in the world have 120,000 cone cells per square millimetre and can spot a red flower from almost a kilometre away.

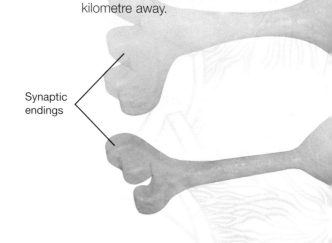

Synaptic endings

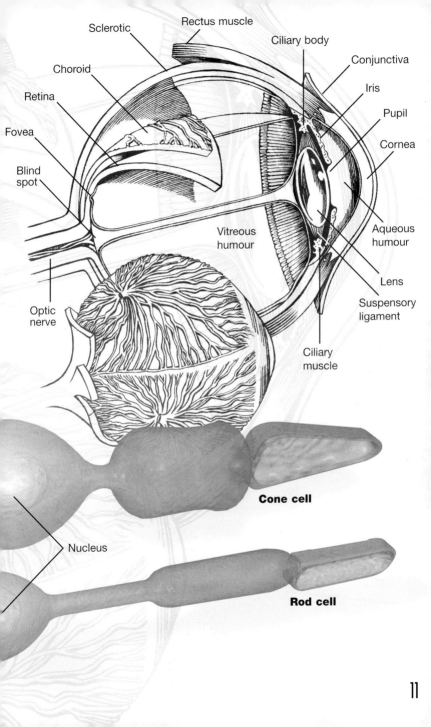

Rectus muscle

Sclerotic

Ciliary body

Conjunctiva

Choroid

Iris

Retina

Pupil

Fovea

Cornea

Blind spot

Aqueous humour

Vitreous humour

Lens

Optic nerve

Suspensory ligament

Ciliary muscle

Cone cell

Nucleus

Rod cell

11

The furthest things we can see

All the stars you can see in the sky are part of a cloud of stars, our galaxy, the Milky Way. Even the nearest stars are several light years away. The nearest star to the Sun, called Alpha Centauri, is 4.29 light years away, but it looks very faint because it's naturally dim. The brightest star in the sky, called Sirius, is also quite close: 8.7 light years away (or 550,000 AU, more than half a million times further from us than the Sun is). It's naturally much brighter than Alpha Centauri, so it looks very bright to us even though it's roughly twice as far away as Alpha Centauri.

The Milky Way is like an island in space. There are many other island galaxies beyond the Milky Way. Two small galaxies, called the Large and Small Magellanic Clouds, are companions to our Milky Way, and can easily be seen in the night sky in the Southern Hemisphere. The nearest large galaxy is called the Andromeda Galaxy. It's about 2 million light years away, and on a very dark night away from city lights it's just possible to see it as a faint patch of light in the sky. This is the furthest thing anyone can see without a telescope.

The stars like dust

On a dark night you can see the Milky Way as a band of light across the sky. Nobody knew what the Milky Way was until Galileo looked at it through a telescope at the beginning of the 17th century and saw that it's made of millions and millions of stars. Without a telescope, even on the darkest night, with no Moon and no clouds, in the middle of the countryside, you can see only about 2,000 stars. These are all quite close to the Sun, by astronomical standards. By counting the number of stars seen in a small patch of the Milky Way with a telescope and multiplying by the size of the whole Milky Way, astronomers calculate that there are two or three hundred billion stars in the Milky Way, all more or less like the Sun.

Seen from outside, the Milky Way Galaxy would look like a huge fried egg, with a bulge of stars in the middle corresponding to the yolk. The whole galaxy is about 100,000 light years across, but only about 1,000 light years thick, outside the bulge. The Sun is about two-thirds of the way out from the centre. It's way out in the galactic suburbs and takes 225 million years to travel once around the centre of the galaxy, moving at a speed of 250 km per second, nearly a million kilometres per hour.

Galileo

The furthest things in the Universe

With the aid of large telescopes, astronomers can look far out into the Universe, beyond the edge of our Milky Way Galaxy. As far as those telescopes can probe, we see other galaxies. Most galaxies are grouped together in systems called clusters. A cluster may contain just a few galaxies, or many thousand galaxies, some smaller than the Milky Way, which is an average-sized galaxy, and some bigger. It's hard to see galaxies that are far away, because the distance makes them look very faint.

31,000 light years away:

The fast-spinning Millisecond Pulsar.

The light we see from this star left when humans were colonising Europe.

2,000 light years away:

The Butterfly Star Cluster in the constellation of Scorpius.

The light we see from these stars left just after the Great Wall of China was built.

Because light takes time to cross space, when we look at a galaxy far away we see it as it was long ago, when the light set out on its journey. So a telescope is a kind of time machine – and so is the human eye, which can also look into the past, although not as far. The light from the Andromeda Galaxy has taken 2 million years to reach us, so we see the stars in that galaxy as they were 2 million years ago. This is called the 'lookback time'. The most distant galaxies seen by the Hubble Space Telescope are more than 10 billion light years away from us. The lookback time is more than 10 billion years, so we can see what the Universe was like 10 billion years ago, 5 billion years before the Earth itself came into existence.

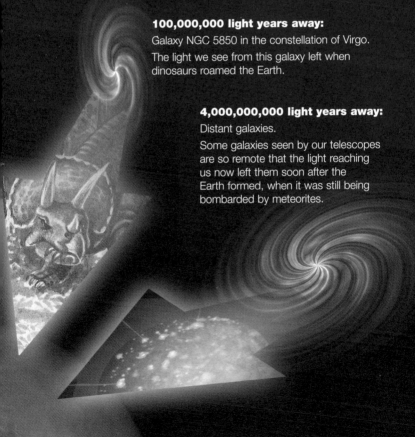

100,000,000 light years away:

Galaxy NGC 5850 in the constellation of Virgo.

The light we see from this galaxy left when dinosaurs roamed the Earth.

4,000,000,000 light years away:

Distant galaxies.

Some galaxies seen by our telescopes are so remote that the light reaching us now left them soon after the Earth formed, when it was still being bombarded by meteorites.

The power of numbers

Very big numbers are sometimes called 'astronomical' numbers, because so many of the numbers that come into astronomy are so big. It's very tedious to write out all these numbers in full, or even to try to give them names. For example, ten billion is 10,000,000,000, and ten billion billion is 10,000,000,000,000,000,000. Scientists have a shorthand way of writing such numbers. The number of 0s is called the 'power of ten' for that number, and they write ten billion as 10^{10} and ten billion billion as 10^{19}. This is just like the shorthand way of writing 100 as 'ten squared' or 10^2, or 9 as 'three squared' or 3^2. The power is also sometimes called the exponent.

Any number can be written in this way, not just ones that are pure powers of ten. For example, 3,456 can be written as 3.456 x 1000, or 3.456 x 10^3. This isn't any shorter for such a small number, but it can be very useful when dealing with astronomical numbers. There's one thing, though, you have to watch out for with this kind of shorthand scientific notation. When you multiply together two numbers written down in this way, you have to add up the powers. So 10^9 x 10^{21} is 10^{30}. And 10^{18} is *not* twice 10^9, it is 10^9 times bigger – that is, a *billion* times bigger!

House: **10 m = 1 x 10^1 m**

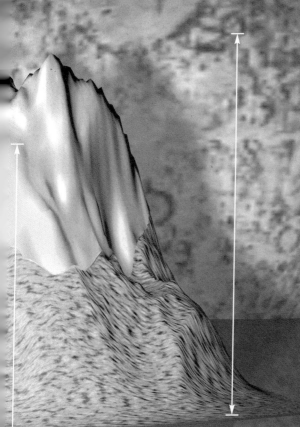

Mount Everest: **8,850 m = 8.85 x 10³ m**

Eiffel Tower: **300 m = 3 x 10² m**

Illustrations not to scale

The Sun is a star

109 Earths

All the stars in the sky are suns, and they all generate their own light. We can see the world around us thanks to our Sun, but we can see the stars in the sky thanks to their own light. The Sun is an ordinary star. Some stars are bigger and brighter than the Sun, and some are smaller and fainter than the Sun. The only special thing about the Sun is that it's the nearest star to us. That's why it looks so big and bright compared with other stars.

Compared with the Earth, the Sun *is* big. It would take 109 planets the size of the Earth, side by side, to stretch across the diameter of the Sun. That means that the volume of the Sun is more than a million times the volume of the Earth. The diameter of the Sun is 1.392 million kilometres, or 1.392×10^6 km.

Mass and weight

The amount of stuff in something is called its mass. A one kilogram bag of sugar has a mass of 1 kilo. The *mass* of your body is the same wherever you are in the Universe (you always have the same amount of stuff in your body), but *weight* is the force experienced by your body as a result of gravitation, and so depends on the gravitational pull of the planet you are on. If gravity is stronger, you weigh more, even though you have the same mass. But on the Moon, you would weigh only one-sixth as much as you do on Earth.

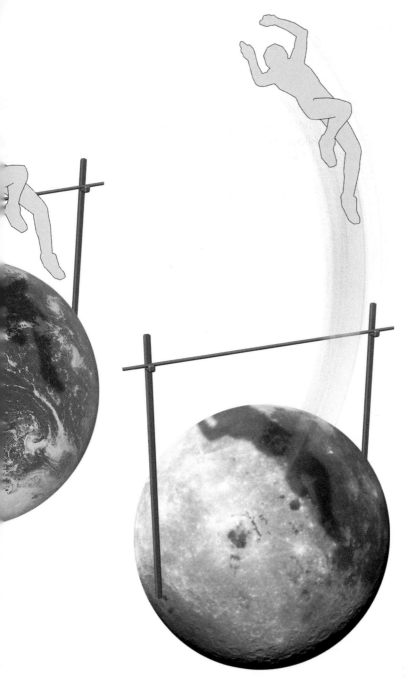

At the heart of the Sun

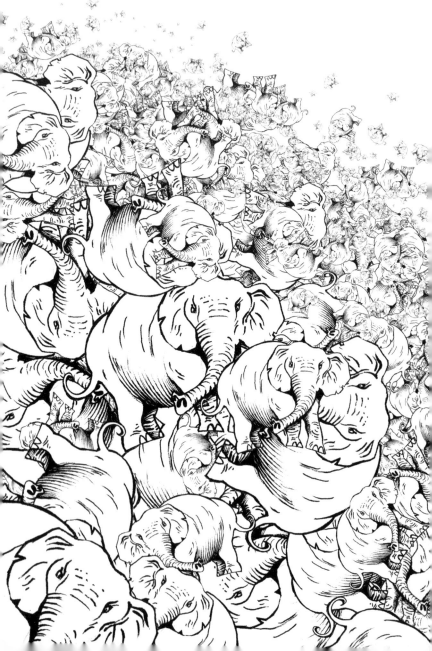

The Sun has a mass of 1.9891 x 10^{30} kilos (about 330,000 times the mass of the Earth), and it radiates 3.83 x 10^{20} megawatts of power into space. If the Sun were entirely surrounded by a shell of ice 2 kilometres thick and touching the surface of the Sun, the ice would be entirely melted in about 50 minutes. And this amount of energy is pouring out of the Sun all the time.

The energy which keeps the Sun shining is released in the heart of the Sun, where nuclear reactions convert mass into energy in line with Albert Einstein's famous equation $E = mc^2$, where E is the energy released from a mass m, and c is the speed of light. In order to keep the Sun shining so brightly, 5 million tonnes of mass are converted into pure energy and radiated away each second. An average elephant has a mass of about 6 tonnes, so that's the equivalent of 833,333 elephants per second! This means that the Sun is losing 5 million tonnes of its substance every second. But the Sun is so big that even though it has been burning up its mass at this rate for 5 billion years, it has still converted only a few millionths of its mass into energy so far.

Packing it in

The density of something tells you how much stuff is packed into a certain space. A kilo of water always has a mass of one kilo, but if it's in the form of steam, it's spread out and has a low density.

The temperature at the Sun's core is 15 million degrees Celsius and the pressure at the heart of the Sun is 300 billion times the pressure of the atmosphere at the surface of the Earth. In the core where the reactions take place, the density of the Sun is 12 times the density of solid lead. That's 160 times the density of water!

conds

10 million years

If light could fly out from the centre of the Sun in a straight line, it would reach the surface in just 2.5 seconds. But because the density is so great, particles of light (photons) from the heart of the Sun are bounced around like balls in a pinball machine for 10 million years before they work their way to the surface.

Oil is less dense than water, and so floats on the surface after an oil spill.

Red giants

The Sun at its present size

The largest stars in the galaxy are called Red Giants, because they are big and red. All ordinary stars like the Sun become Red Giants for a time when they become middle-aged. This will happen for the Sun in about 5 billion years from now. It happens because the star has used up some of its nuclear fuel. This makes the middle of the star shrink and get hotter, which starts a second wave of energy production. But even though the middle of the star has shrunk, the extra heat makes its outer layers swell up.

A Red Giant may be as much as 100 times bigger in diameter than the Sun is today – that is, about 10,000 times wider than the Earth. If the Sun expanded to become a Red Giant today, it would be so big that it would actually swallow up the Earth. But by the time it does become a Red Giant, it will have lost about a quarter of its mass by blowing gas away into space. Even so, it will still expand beyond the orbit of Mercury, the innermost planet. The heat from the huge red Sun will boil the oceans, and winds of gas from it will strip away the atmosphere of the Earth, as well as the atmospheres of Venus and Mars. The surviving inner planets of the Solar System will be nothing more than dead cinders of rock. But we still have 5 billion years to figure out where to go when this happens!

Red giant

29

White dwarfs

White Dwar

The Sun at its
present size

Sun	White Dwarf
○	○
Red Giant	Sun

At the very end of its life, after its time as a Red Giant, a star like the Sun runs out of fuel and cannot generate heat in its heart any more. As it cools down, it shrinks into a ball of star stuff about the size of the Earth. But even though its radius is only 0.01 of the radius of the Sun, this ball still contains as much matter as there is in a star, so it's very, very dense. A single thimble-full of this dense star stuff would have a mass of about 1 tonne, a million times the density of water. At first, these dense stars glow because of the leftover heat from their youth; they may be as hot as 10,000 degrees Celsius, and are called White Dwarfs. Eventually, they cool down and fade away, to become Black Dwarfs.

The pull of gravity at the surface of a White Dwarf is more than 10,000 times as strong as the pull of gravity at the surface of the Earth. A person who weighs 75 kilos on Earth would weigh 750,000 kilos, or 750 tonnes, on the surface of a White Dwarf, and would be crushed by their own weight.

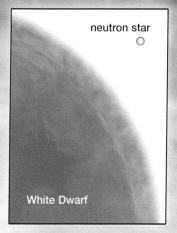

White Dwarf

neutron star

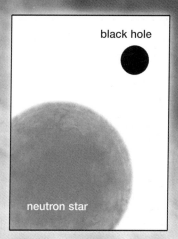

black hole

neutron star

Neutron stars, supernovae and hypernovae

If a star starts out in life with much more mass than the Sun, it can't settle down quietly as a White Dwarf. When it runs out of fuel, it starts to collapse under its own weight, but this turns huge amounts of gravitational energy into heat. The heat makes the star explode as a supernova. Temperatures inside the supernova reach as high as 48 billion degrees Celsius. This incredible heat makes the dying star shine as brightly as a whole galaxy for a few days before it flings off large amounts of matter into space. If the amount of matter left behind is between about 1.4 and 3 times the mass of our Sun, its own weight squeezes it so hard that even atoms (see page 146 onwards) are squashed. This forces electrons to combine with protons to make neutrons (see pages 168–9). The result is a ball of neutrons. Even though only about 10 km across, about the size of Mount Everest, this ball of neutrons would contain twice as much mass as our Sun. A thimble-full of a neutron star would contain as much mass as all the people on Earth put together. The pull of gravity at the surface of a neutron star is a hundred billion (10^{11}) times stronger than on Earth, so the surface is smoothed into an almost perfect sphere.

The biggest explosions ever seen release hundreds or thousands of times more energy than supernovae. They are called hypernovae, and most occur in galaxies far, far away across the Universe. The energy released in a hypernova explosion is thought to be the result of a very massive star collapsing completely to become a black hole. If a hypernova occurred as close to us as 2,000 light years, it would look twice as bright as the Sun in ordinary light, but would 'shine' 2,000 times brighter than this in gamma rays (see pages 84–5), killing most life on the surface of the Earth.

Tiny amount of neutron star matter

Black holes

If a star has more than three times as much mass as the Sun left over at the end of its life, it can't settle down as a White Dwarf or even as a neutron star. It vanishes into a point, collapsing completely under its own weight, and is crushed out of existence entirely. But we can never see this point (called a singularity) because gravity stops anything at all escaping, including light. The place where gravity stops light escaping is called the event horizon, because you can't see any 'events' going on inside it. For that same reason we can't illustrate a singularity here, but can only give an impression of its effects. Because light can't escape from such a collapsed star, it's called a black hole.

Anything will make a black hole if it's squeezed hard enough. But the more stuff you start with, the less you have to squeeze. The Earth would become a black hole if it were squeezed to the size of a large pea. The Sun would become a black hole if it were squeezed to a ball 2.9 km across. But if you had a 'drop' of water as big across as our Solar System, it would become a black hole without any squeezing. Astronomers detect black holes because matter falling into them gets very hot and radiates X-rays (see pages 84–5) before it disappears. There are many very massive black holes in the centres of galaxies. In 1994, observations with the Hubble Space Telescope showed that there is a disc 500,000 light years across orbiting at speeds of 2 million km per hour around a black hole. This black hole has a mass 3 billion times the mass of our Sun and is in the galaxy M87, 50 million light years from Earth.

Black holes

If a star has more than three times as much mass as the Sun left over at the end of its life, it can't settle down as a White Dwarf or even as a neutron star. It vanishes into a point, collapsing completely under its own weight, and is crushed out of existence entirely. But we can never see this point (called a singularity) because gravity stops anything at all escaping, including light. The place where gravity stops light escaping is called the event horizon, because you can't see any 'events' going on inside it. For that same reason we can't illustrate a singularity here, but can only give an impression of its effects. Because light can't escape from such a collapsed star, it's called a black hole.

Anything will make a black [hole if squ]eezed hard enough. But the more stuff you start [with the harder to] squeeze. The Earth would become a b[lack hole if crushe]d to the size of a large pea. The Sun wou[ld become one if it we]re squeezed to a ball 2.9 km acros[s. There could be a black h]ole as big across as our Solar System [if it formed naturally, be]cause without any squeezing. Astron[omers detect very massive b]lack holes pages 84–5) before it disappears. There are m[any very massive b]lack holes in the centres of galaxies. In 1994, observations with the Hubble Space Telescope showed that there is a disc 500,000 light years across orbiting at speeds of 2 million km per hour around a black hole. This black hole has a mass 3 billion times the mass of our Sun and is in the galaxy M87, 50 million light years from Earth.

Across the Solar System

An Orrery

A clockwork model of our planetary system.

The Oort Cloud

Far beyond Pluto, the Oort Cloud contains about a hundred billion individual comets.

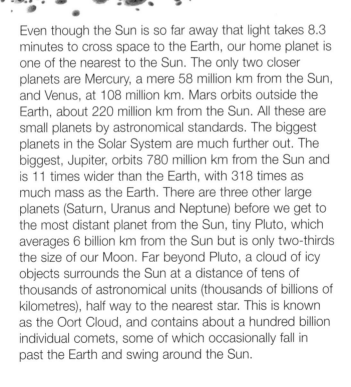

Even though the Sun is so far away that light takes 8.3 minutes to cross space to the Earth, our home planet is one of the nearest to the Sun. The only two closer planets are Mercury, a mere 58 million km from the Sun, and Venus, at 108 million km. Mars orbits outside the Earth, about 220 million km from the Sun. All these are small planets by astronomical standards. The biggest planets in the Solar System are much further out. The biggest, Jupiter, orbits 780 million km from the Sun and is 11 times wider than the Earth, with 318 times as much mass as the Earth. There are three other large planets (Saturn, Uranus and Neptune) before we get to the most distant planet from the Sun, tiny Pluto, which averages 6 billion km from the Sun but is only two-thirds the size of our Moon. Far beyond Pluto, a cloud of icy objects surrounds the Sun at a distance of tens of thousands of astronomical units (thousands of billions of kilometres), half way to the nearest star. This is known as the Oort Cloud, and contains about a hundred billion individual comets, some of which occasionally fall in past the Earth and swing around the Sun.

Going to extremes

There is a huge range of conditions to be found on the planets of the Solar System. Pluto is the coldest planet, with a temperature of –220 degrees Celsius, cold enough to be covered in ice made of frozen methane, the gas used in camping stoves on Earth. The temperature on Mercury goes up only to about 200°C, even though it's so close to the Sun, because it has no atmosphere to trap heat. But Venus has an atmosphere so thick that the pressure at its surface is 90 times the pressure of the air at sea level on Earth, and this traps so much heat (by the greenhouse effect) that the temperature soars to about 600°C.

Jupiter is almost entirely a ball of gas, surrounding a tiny rocky core. This core is still at a temperature of about 20,000°C, heat left over from the time when the planet formed, and the pressure on it is several million times the pressure of air on Earth at sea level. This pressure is so great that the rocky core is surrounded by a layer of liquid hydrogen.

The structure of Jupiter

Due to the enormous pressures, the temperature of Jupiter is believed to increase by 0.3 °C for every kilometre of depth.

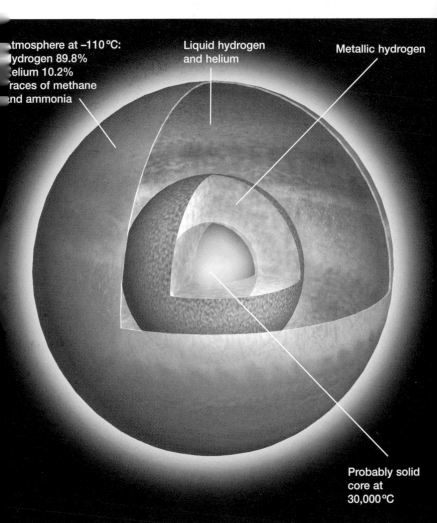

Atmosphere at –110 °C:
Hydrogen 89.8%
Helium 10.2%
Traces of methane
and ammonia

Liquid hydrogen and helium

Metallic hydrogen

Probably solid core at 30,000 °C

The death of the dinosaurs

The comets that fall into the inner part of the Solar System don't always miss the Earth. About 65 million years ago, a lump of ice about 10 km across, travelling at a speed of about 50 km/sec (180,000 km/hour), hit the Earth in a shallow sea near what is now Mexico. The impact released as much energy as the explosion of about a hundred million megatonnes of TNT. The energy released would have blasted lumps of molten rock into the sky. They would have fallen back to the surface all over the Earth, spreading the heating out to make the equivalent of 10 kilowatts per square metre for several hours. The entire surface of the Earth would have been as hot as the inside of an oven roasting a chicken. Then, dust from the explosion would have stayed in the air for years, blocking sunlight and killing the green plants that depend on it to survive, and plunging the planet into an ice age. All of this, and other side effects from the impact, brought an end to the age of the dinosaurs.

The shaking earth

Our planet doesn't always need help from outside to produce explosions felt around the world. In 1883, the volcanic island of Krakatau, near Java, exploded with the force of 200 megatonnes of TNT, killing 36,000 people in the immediate vicinity and more than 50,000 when a tidal wave 30 metres high, caused by the explosion, swept over the nearby islands. The explosion was heard nearly 5,000 km away in Mauritius, and it shot 20 cubic kilometres of dust into the upper atmosphere, producing coloured sunsets around the world for months. The temperature over the whole Northern Hemisphere fell by 0.5°C in the 1880s, as this dust acted as a kind of sunshield.

The most famous earthquake of all time, which hit San Francisco in 1906, amazingly killed only a few hundred people. But the fire which raged for three days after the 'quake destroyed 30 schools, 80 churches and the homes of 250,000 people, doing damage worth $400 million. In 1923, an earthquake almost exactly the same size hit Japan, and this time more than 140,000 people were killed in fires that raged through Tokyo and Yokohama.

All of this happens because the 'solid Earth' isn't solid. The Earth's crust is made up of several pieces, called tectonic plates, which fit together like a spherical jigsaw puzzle. Volcanoes and earthquakes are common where the edges of these plates rub together.

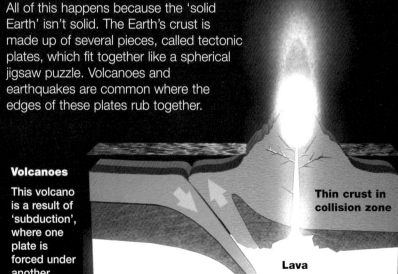

Volcanoes

This volcano is a result of 'subduction', where one plate is forced under another.

Thin crust in collision zone

Lava

The structure of Earth

To get some idea of how thin the Earth's crust is, compare it to the thickness of the skin on an apple.

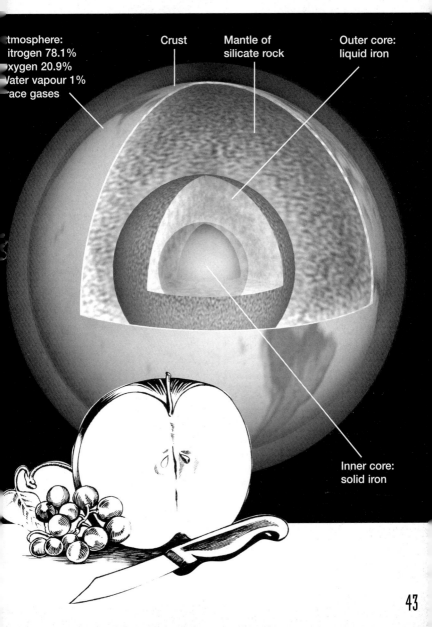

Atmosphere:
Nitrogen 78.1%
Oxygen 20.9%
Water vapour 1%
Trace gases

Crust

Mantle of silicate rock

Outer core: liquid iron

Inner core: solid iron

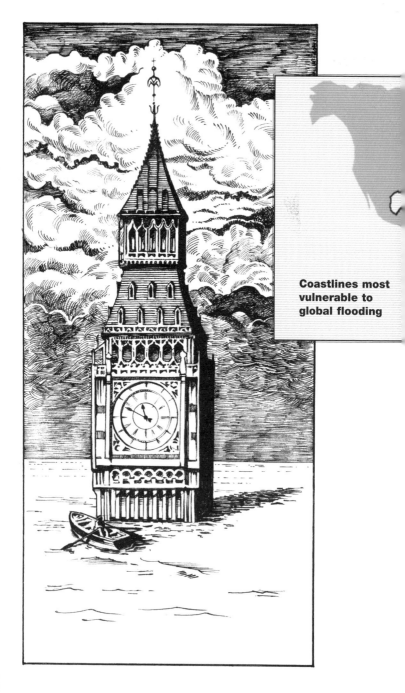

**Coastlines most
vulnerable to
global flooding**

More extremes

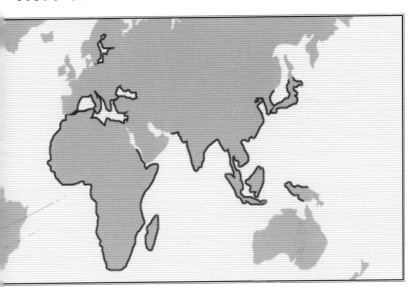

In the hot deserts of the world, daytime temperatures often climb above 35°C (95 degrees Fahrenheit), with an official record of 58°C (136.4°F) in Azizia, Libya. If you found a big flat stone out in the direct sunlight at that temperature you would be able to fry an egg on it. But because of the clear skies, night-time temperatures in these same desert regions fall below 5°C, and frosts are not unknown. The coldest temperature ever recorded, however, was –91°C (–131.8°F) in the depths of the Antarctic winter.

Such cold temperatures ensure that the entire continent of Antarctica is covered by ice. This permanent ice cap, grounded on land, covers an area of 15,500,000 square km and ranges from 1.7 to 2.2 km in thickness. The total volume of fresh water contained in the ice cap is 70% of all the world's fresh water: 30 million cubic kilometres or 3×10^{19} litres, enough to fill 1.3×10^{17} ordinary domestic bath tubs. The Antarctic continent is also surrounded by a fringe of floating sea ice. This has been shrinking in recent years, because the world is getting warmer. In 2002, a huge chunk of ice called the Larsen B ice shelf broke away. It was 3,250 square km in area, about the size of Yorkshire.

The largest living thing

In the Malheur National Forest in the Blue Mountains of Oregon there is a species of fungus called *Armillaria ostoyae,* more commonly known as the honey mushroom, that covers 891 hectares (2,200 acres) of land. On a piece of land that size you could fit 1,220 football pitches. But this honey fungus has been taking its time growing to such an incredible size, and it has probably been there for 2,500 years.

The tallest tree is a *Sequoia sempervirens,* usually called a giant coast redwood. There is one in Montgomery State Reserve in California which has been growing for over 1,000 years and has reached a height of 112 metres.

Redwoods are the most massive trees as well as the tallest. The most massive ever recorded blew over in a gigantic storm in 1905 and weighed 3,688 tonnes, about the same as 20 blue whales or 600 elephants. The largest living tree is a giant sequoia called 'General Sherman'

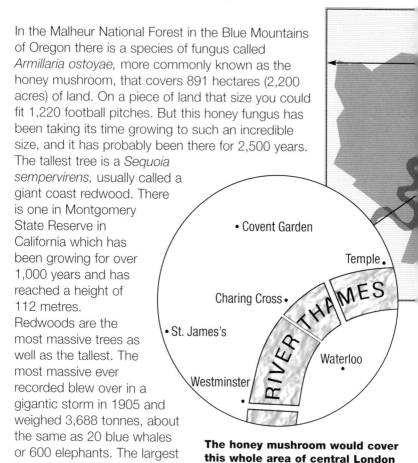

The honey mushroom would cover this whole area of central London

which is growing in the Sequoia National Park in California. If it were cut down and used for timber there would be enough wood to make 5 billion matchsticks. The biggest living creature is *Balaenoptera musculus,* the blue whale. The biggest one ever caught weighed 190 tonnes, as much as about 30 elephants.

45 km

Area covered by honey mushroom compared to the area of London

A bug's life

The smallest living things are bacteria and viruses. A bacterium is a single little blob of life, a bag of complex chemicals held inside a skin, like a tiny water-filled balloon, and called a cell. Complex life forms like plants and animals are made up of many cells working together. A virus is a package of protein-covered chemicals that breaks into living plant or animal cells, but does not have its own cell wall.

Some bacteria cause diseases, but many are harmless, and some are actually helpful. Bacteria in your stomach help you to digest food. Shared out among all the 6.5 billion people on Earth, there are about 4×10^{23} of these gut bacteria. If you had that many pennies in a pile, it would be 100,000 light years tall. That's about six thousand billion bacteria per person.

Bacteria are part of a family of living things called prokaryotes. All prokaryotes are made of just a single cell. Although individual prokaryotes are too small to be seen, all of them put together would weigh as much as all the living things we can see, like trees and whales and people. One glass of pasteurised milk contains 6,000,000 bacteria, half a pound of raw minced beef contains 224,000,000 bacteria, and one drop of human saliva from a healthy person contains 150,000,000 bacteria.

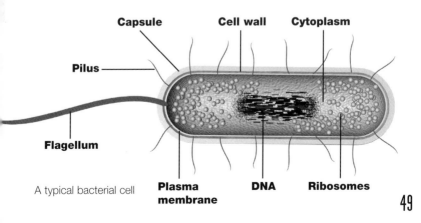

Capsule **Cell wall** **Cytoplasm**

Pilus

Flagellum

A typical bacterial cell

Plasma membrane **DNA** **Ribosomes**

Life that feeds on life

Viruses are a strange form of parasitic life. They are totally dependent on cells. But viruses are very, very much smaller than cells and there are billions of viruses waiting to be breathed in around you all the time. Inside a cell, they hijack the machinery of life and use it to make more viruses, killing the cell in the process. Outside the cell, they exist in suspended animation as tiny particles only a few billionths of a metre across.

When someone with flu sneezes, 100,000 viruses are spread over a distance of 9 metres, the length of a London bus. The sneeze travels at more than 160 kilometres (100 miles) per hour, so it covers 10 metres in a quarter of a second.

How plagues spread

Bacteria are found all over the Earth: on the ground, in seas, lakes, rivers and ice, in people, plants and animals. Some bacteria are useful and break down waste, but some cause dangerous diseases. A bacterium called *Clostridium botulinum* is so deadly that just 450g of it could kill the entire population of the world. One bacterium is 1,000 times smaller than an animal cell, so small that it can be seen only with an electron microscope.

Before effective antibiotics were invented, plagues swept uncontrollably through huge areas, killing much of the population. In the 14th century, a plague called the Black Death, which was caused by a bacterium carried by fleas that lived on rats, swept through Europe, killing more than 25,000,000 people – a quarter of the population. In March 1918, a dangerous form of influenza originated in the USA and was carried all over the Earth in about four months, killing at least 21,000,000 people before it was brought under control.

A deadly virus

AIDS (Acquired Immune Deficiency Syndrome), a modern-day plague which has already affected 35,000,000 people worldwide and is responsible for the deaths of around 3,000,000, is caused by a virus which stops the body's natural defences from working. This defence breakdown enables other viruses and bacteria to attack the body and so cause disease and death.

But viruses and bacteria aren't really doing anything different from animals like wolves and lions. They're hunting for food to eat, and not 'trying' to make us ill. It's just our bad luck that the cells in our bodies are made of the food they need.

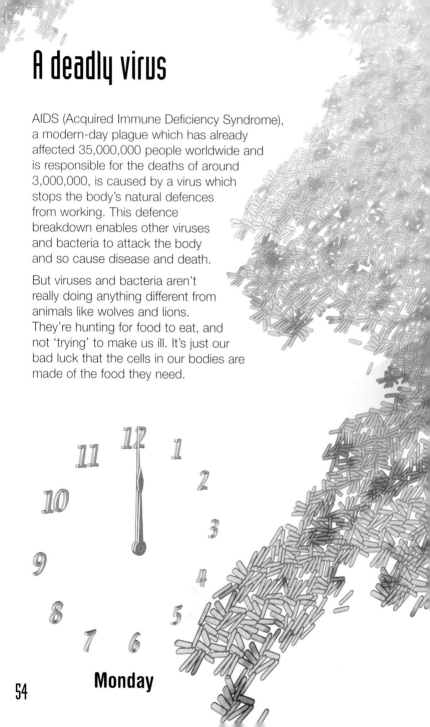

Monday

Tuesday

24 Hours

When people get ill, they sometimes say that they have 'caught a bug'. But it won't be just one 'bug', or bacterium, that makes them ill. Bacteria reproduce extremely quickly. Each bacterium is a single cell, and provided it has a supply of food it can quickly grow enough to divide in two, making two bacteria each just like the parent. For some bugs, the population doubles every ten minutes. This means that if you start with a single bacterium, after ten minutes you have two, after twenty minutes you have four, after thirty minutes you have eight, and so on. This is exponential growth (see pages 88–9). In less than two hours, each original bacterium has more than a thousand descendants. After six hours, each bacterium has billions of descendants, and in less than 24 hours a single bacterium can have more than four quadrillion (4,000,000,000,000,000) descendants. If they have the ideal conditions in which to grow, the population can increase until there are 10 billion bacterial cells in each millilitre.

The web of life

The basic unit of life is the cell. A cell is a bag of complex chemicals in a watery soup, held together behind a skin called the cell membrane. All the chemistry of life goes on inside the cell.

The cell of a single bacterium occupies a volume of about 1 cubic micrometre. So it would take 10^{18} bacteria (that is, a billion billion bacteria (1,000,000,000,000,000,000)), packed side by side, to fill a volume of 1 cubic metre (a thousand litres) such as a packing case measuring 1 metre on each side. The cells of more complicated organisms, such as people or carrots, are much bigger, two or three thousand times the volume of bacterial cells. But even so, the human body contains about a hundred thousand billion cells, which is about 500 times the number of bright stars in the Milky Way Galaxy. People are much more complicated than galaxies, and animals like us are the most complicated things we know of in the entire Universe.

In your blood

There are two important kinds of cell in your blood. White cells attack invaders, such as bacteria, and destroy them. Red cells carry oxygen around from your lungs to wherever it's needed. But each red cell has a relatively short lifetime. They're made in your bone marrow, at a rate of 140,000 per minute (that's 8,400,000 per hour). Each red cell flows around in the blood for a few months, doing its work, before it gets worn out. The used-up red cells are broken down in the liver (at the same rate that they're being made by bone marrow) and their components are recycled.

Blood is rapidly pumped around the body by the heart. It travels through the arteries, which carry it to the body tissues such as muscles and organs, and then through the capillaries, which are a fine network of smaller vessels contained within the tissues. Finally, the veins, which are similar in size to the arteries, return the blood to the heart.

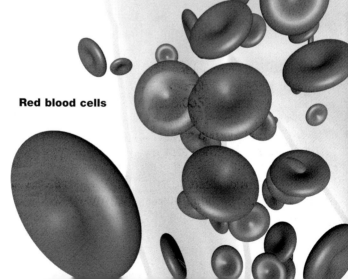

Red blood cells

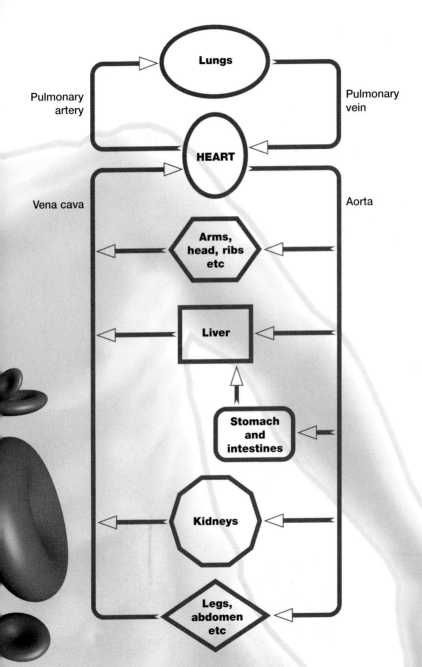

Heartbeat

During your entire lifetime, your heart will beat about 1,500,000,000 times. The strange thing is, exactly the same would be true if you were a mouse or an elephant. All land mammals live for roughly the same number of heartbeats. A small mammal, like a mouse, loses heat quickly because its volume is small compared with its surface area. So it uses up its food for energy very quickly, and has a rapid heartbeat but a short life. A large mammal, like an elephant, loses heat slowly because its surface area is small compared with its volume. So it uses up its food for energy very slowly, and has a slow heartbeat but a long life. We are roughly in the middle.

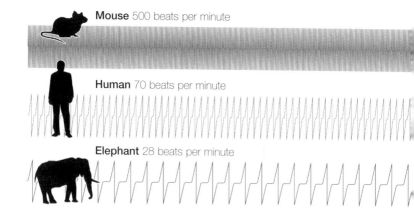

Mouse 500 beats per minute

Human 70 beats per minute

Elephant 28 beats per minute

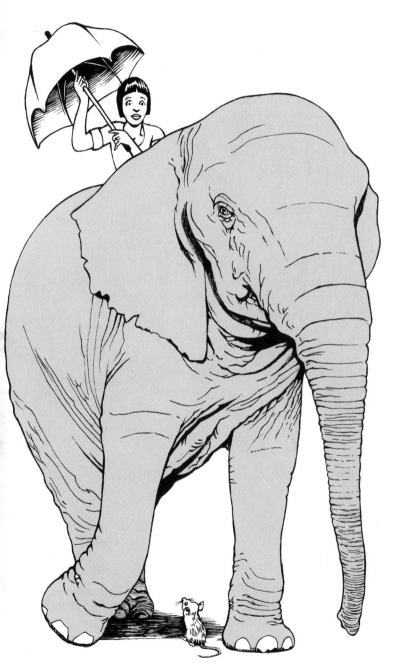

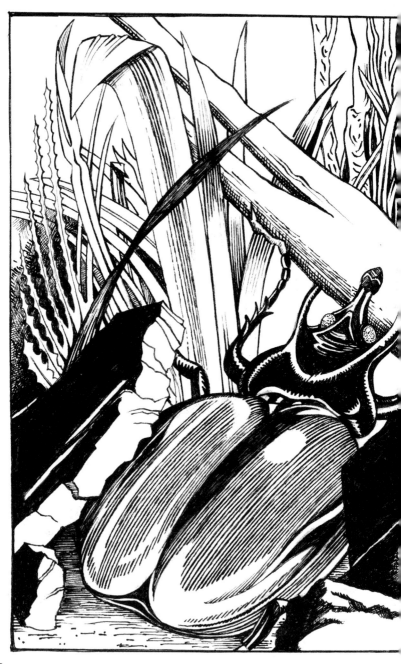

The biggest isn't always the strongest

An African elephant can lift almost 3,000 kg, about 25% of its body weight. This means that it can move some tree trunks as easily as you would pick up your rucksack. But the strongest creature in proportion to its size is a rainforest beetle that you could hold in the palm of your hand.

There are about 300 different species of rhino beetles (scarabs). Most live in tropical countries, but there are some in North America. As beetles go, they look quite scary because they are among the world's largest, and they have horns on their heads very much like a rhinoceros. But these beetles are vegetarians and are totally harmless to people. Only males have 'rhino' horns, and sometimes when two males are fighting over who gets the best feeding site they lock horns in combat. Female rhino beetles are attracted to the males who have the best feeding sites, so fighting male rhinos are clashing over who gets the food and the girls!

A rhino beetle's horns are also useful for digging through the heavy plant litter on the jungle floor, and it can move obstacles much bigger than itself out of the way. When it needs to, a male rhino beetle can support up to 850 times its own weight on its back. That's the same as if a man could carry 76 family cars around with him at the same time!

The brainy ape

The most complicated organ in the human body is the brain. This means that it's the most complicated thing we have discovered in the known Universe. Some animals, like whales and elephants, have larger brains than us, but they also have much larger bodies. The human brain weighs about 1.4 kilos, which is much bigger in proportion to the size of the body it lives in, and it's much more complicated even than these larger brains.

An adult human brain contains about 100,000 million special cells called neurons which transmit and receive information. Neurons are cells which pass information from one to another like messages going along telephone wires. The thinking bits of your brain are the electrically excitable neurons. A neuron can trigger a reaction far away, and this communication is called the nerve impulse. The speed of a nerve impulse can be as fast as 400 kilometres an hour. Neurons are linked to one another by about 1,000,000,000,000 separate connections.

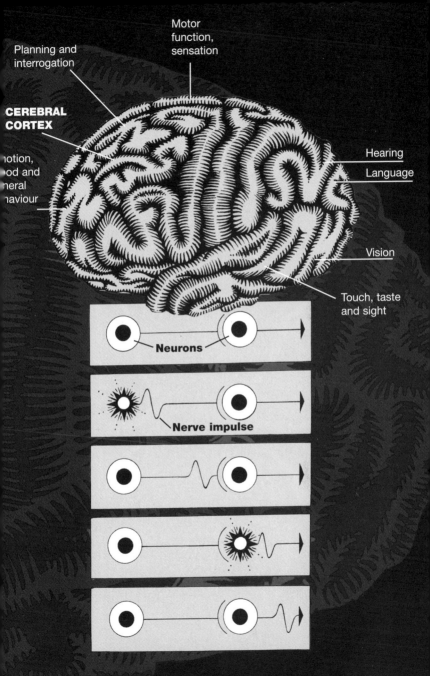

Motor function, sensation

Planning and interrogation

CEREBRAL CORTEX

...otion, ...od and ...neral ...haviour

Hearing

Language

Vision

Touch, taste and sight

Neurons

Nerve impulse

Working together

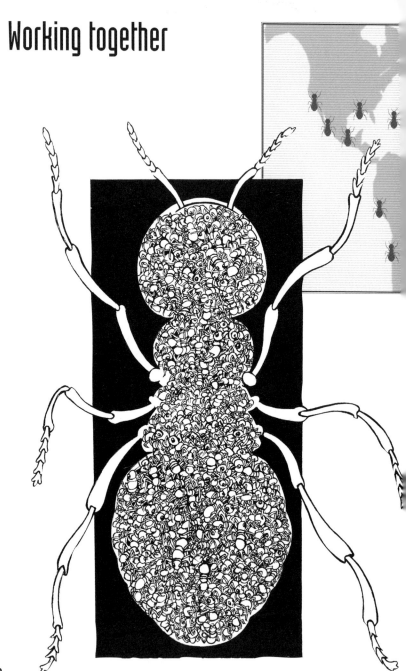

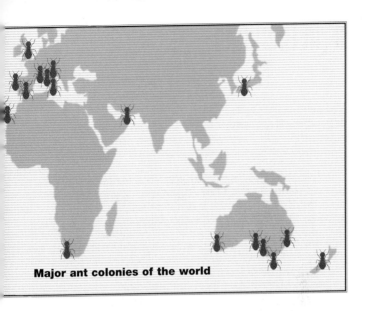

Major ant colonies of the world

Rather like the way cells work together to make up a living body, in some species many individual creatures work together to make up a single living unit. The bees in a hive, for example, are much more dependent on each other than the people who live in a big city. The queen lays eggs, her attendants look after them, and workers find food. But none of them can do the other one's job. A swarm of bees might contain 40,000 or 50,000 individuals, but it's really just one living system. Ants and termites have developed this lifestyle in much greater numbers. They are so successful that it's estimated there are a million billion ants alive on Earth today – 168,000 ants for every human being.

The largest co-operative unit of individual living organisms known on Earth is a colony of Argentine ants which stretches from the northwest of Italy around the Mediterranean coasts of France and Spain, and up the Atlantic coastline of Spain and Portugal for a distance of 5,953 km. Argentine ants co-operate within closely related colonies, but fight fiercely with members of other colonies. The ants in this super-colony are all closely related, and don't fight one another. It's estimated that this single colony contains thousands of billions of ants, at least ten times as many as the number of stars in the Milky Way.

Blowin' in the wind

Not all plants have flowers or seeds. Some reproduce by spreading spores. Spores are very small. Ferns reproduce by spores and you can see clusters of them, called sori, underneath their leaves. When the spores are ripe, these sori split open and the spores blow away in the wind. A big fern can produce a hundred million – 100,000,000 – spores a year, but only a few will manage to settle and grow into new ferns. The spores from mushrooms and toadstools are shed by the gills that you can see on the underside of their caps. Mushrooms and toadstools have long stalks because the spore-releasing gills under the cap need to be well above ground so that the spores can drift off easily and spread through the air.

Mushrooms, toadstools and puffballs are the visible parts of fungi, and they hold all of the spores. Puffball fungi are like bags that push out their spores through a hole in the top, a bit like a volcano. An average-sized puffball fungus can spew out 7 million million spores and one as big as a football could produce 10 million million spores. The biggest ever edible fungus was a giant puffball found in Canada which weighed in at 22 kilos. This fungus would have been able to produce billions and billions of spores.

Everybody has two parents and four grandparents. Those parents and grandparents each have two parents and four grandparents. Every time you go back one generation, you double the number of your ancestors. After ten generations, that means you are descended from 1,024 people. If parents are, on average, 25 years older than their children, that means that just 250 years ago there were 1,024 of your ancestors alive. But the same argument holds for every living person on Earth, so *everybody* had 1,024 ancestors alive 250 years ago. And yet, the population of the Earth was much smaller 250 years ago than it is today. What is wrong with this argument?

The numbers are right, but they don't take account of the fact that each ancestor may have lots of children and grandchildren and so on. If you have a sister, you and your sister both have four grandparents, but they are the same four grandparents! If you have a first cousin, you have only six grandparents between you. You really do have about 1,000 ancestors who were alive 250 years ago, but most of those are also the ancestors of your friends, and people you have never met. Everybody is related to everybody else.

Child 1

Child 2

Common
great-
grandparent

Two children sharing one
great-grandparent share at
least 128 of their 1,024
tenth-generation ancestors

Counting descendants

One pair of ancestors can leave many descendants. This puzzled
people like Charles Darwin in the middle of the 19th century.
Elephants breed very slowly, with a gap of about 30 years between
generations. But if each pair of elephants produced just four
offspring that survived to have offspring in their turn, this would still
make the population of elephants double every 30 years. After 25
doublings (750 years) there would be more than 33 million elephants
in the present generation descended from *each* original pair! But
Darwin knew, of course, that there were about as many elephants
around in 1850 as there had been in the year 1100. The reason is
that, on average, each pair leaves only two offspring that survive to
have offspring of their own. In all species on Earth (including
humans, until very recently), many offspring die without leaving
descendants. In 1920 there were about 1.5 million elephant alive in
the world, but today the number is down to around 60,000.
Because the ivory trade, which was responsible for the deaths of a
huge number of elephants, is now banned worldwide, the number of
elephants on Earth is now increasing.

73

Fruitful flies

Fruit flies are only half a millimetre long, but they can show how rapidly populations could increase if they had the opportunity. One fruit fly can lay 500 eggs, which hatch within 24 hours, which develop into tiny maggots that grow up into adults that can lay 500 eggs in their turn. If all the eggs hatched into females (in real life, of course, half of them will be males), starting with one fly, you would have 500 flies after a week, 250,000 flies after a fortnight, 125,000,000 flies after three weeks and 625 hundred million (62,500, 000,000) flies after a month.

Week 4:
62,500,000,000

Week 3:
125,000,000

Week 2:
250,000

Week 1:
500

74

Illustrations not to scale

Quite a mouthful

Every drop of sea water contains plankton, which is the starting point of the sea's food chain. Plankton is made up of minute animals and plants that drift near the surface of fresh or salt water. The plants are called phytoplankton and the animals are called zooplankton. Animal plankton includes single-cell animals, jellyfish and crustaceans. Oceans cover three quarters of the Earth's surface, and so plankton makes up the largest amount of living creatures on Earth.

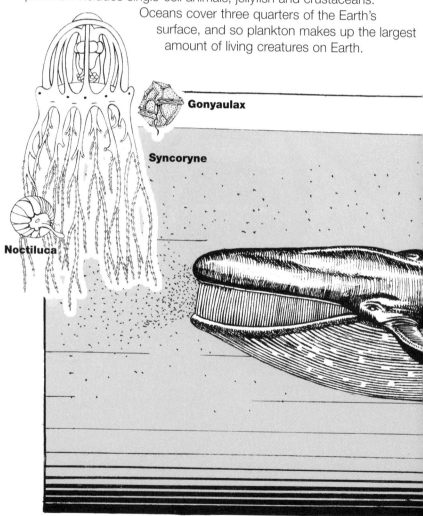

Gonyaulax

Syncoryne

Noctiluca

Most plankton drift along with the ocean currents, but some are amazingly strong swimmers. One type of zooplankton called Dinoflagellates have a light-sensitive spot that draws them up to the surface of the ocean during the light of day, and then they descend back to the ocean depths at night. They can travel 45 metres in 24 hours, which is the equivalent of a human swimming 3,200 km in a day.

Phytoplankton can be found in every ocean and are the most abundant type of plant life on Earth. There are countless billions of zooplankton in the oceans of the world. They feed off phytoplankton, and are then in their turn eaten by small fish, which are eaten by other fish and marine animals.

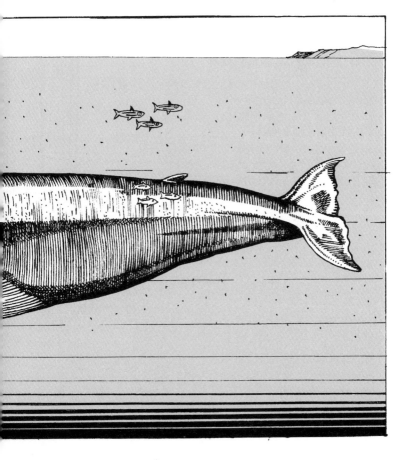

Super computers

The incredible speed and capabilities of modern super computers has revolutionised the speed at which extraordinarily complex calculations can be made. They are used to help predict the weather, as well as for scientific and engineering research.

William Shanks, a 19th-century teacher, took 28 years of his spare time working out the mathematical value of the number pi (the ratio of the circumference of a circle to its diameter; see pages 98–9) to 707 places – a task that would take a super computer a second to complete. To make matters worse, Shanks made an error with his 528th calculation, so he spent all that time on it and still didn't get it right!

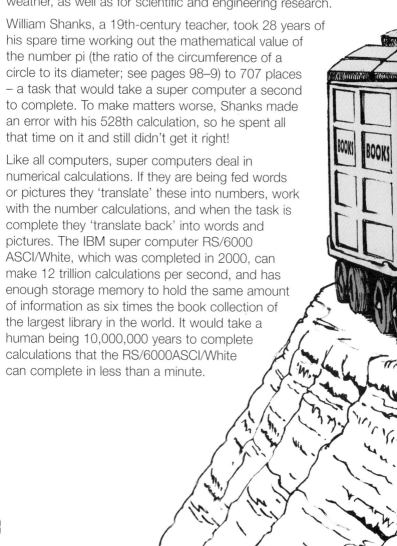

Like all computers, super computers deal in numerical calculations. If they are being fed words or pictures they 'translate' these into numbers, work with the number calculations, and when the task is complete they 'translate back' into words and pictures. The IBM super computer RS/6000 ASCI/White, which was completed in 2000, can make 12 trillion calculations per second, and has enough storage memory to hold the same amount of information as six times the book collection of the largest library in the world. It would take a human being 10,000,000 years to complete calculations that the RS/6000ASCI/White can complete in less than a minute.

Swifter than a speeding bullet

The speed of a bullet shot from a rifle is 3,500 km per hour, and pilots have travelled at speeds that almost match this. Major George Morgan and Captain Eldon Joersz of the US Air Force flew a Lockheed SR.71A aircraft at 3,529.56 km per hour in 1976.

Spacecraft move even faster. The fastest speed at which a human being has ever travelled was reached by Eugene Cernan, Thomas Stafford and John Young, the US crew of the command module of Apollo 10. On Apollo 10's trans-Earth return flight on 2 May 1969, the craft travelled at 39,897 km per hour – an amazing 11.08 km every second.

Apollo 10 may have reached the fastest speed, but it was the crew of Apollo 13 that flew the furthest distance out from the surface of the Earth on 15 April 1970. Fred Haise, Jim Lovell and Jack Swigert, the all-US crew of Apollo 13, travelled 400,171 km out into space.

Gravity makes everything you throw into the air come back down to the ground. Rockets have to reach at least 40,000 km per hour to be at a speed fast enough to escape the pull of the Earth's gravity and go on up into space. The force pushing a rocket as it leaves the Earth has to be bigger than the force of gravity pulling towards the ground. So for rockets to get into space, 40,000 km or more per hour is the 'escape velocity'.

Tsunami – the giant wave

Under-
sea earthquakes cause
vibrations to surge through the
sea at many hundreds of
kilometres per hour. These
may have a wavelength – the
distance between the crests –
as long as 200 km, but a height of
only about 50 cm. A passenger on a
ship wouldn't even notice such a ripple, even
though it travels at 200 metres per second. But as the wave
approaches land, the friction of the sea floor makes it get
steeper and steeper. By the time it hits the shore, it may be
30 metres high – a gigantic wave called a tsunami (Japanese
for 'harbour wave'). Many people call them tidal waves, even
though they have nothing to do with tides. A tsunami can be
so strong that it destroys everything in its path.

On 26 December 2004 an earthquake measuring 9 on the
Richter scale occurred deep in the Indian ocean. It caused a
huge tsunami which raced across the ocean at a speed of
over 800 km per hour to hit coastal areas of Sri Lanka, India,
Indonesia and Malaysia. It even reached the east coast of
Africa over 6000 km away. The damage the tsunami caused
was immense; more than a hundred thousand people were
killed and millions of lives were affected. It destroyed vast
areas of land and many houses and buildings, and even
whole towns, were swept away by the water.

The length of light

Light travels as waves, like ripples on a pond. But the wavelength of light is much smaller than the wavelength of ripples on water. The wavelength of light is related to its colour. Violet light, at one end of the rainbow spectrum, has a wavelength of about 380 billionths of a metre (0.00000038 m, or 3.8 x 10^{-7} m). Red light, at the other end of the spectrum that is visible to our eyes, has a wavelength of about 750 billionths of a metre (7.5 x 10^{-7} m). Light waves are made of

1 nanometre (nm) = one billionth of a metre

X-rays: Between 0.1 and 10 nanometres.

High-energy electromagnetic radiation emitted by hot gas, X-rays are found near black holes and between galaxies. They are used here on Earth by doctors to show the interior of the body.

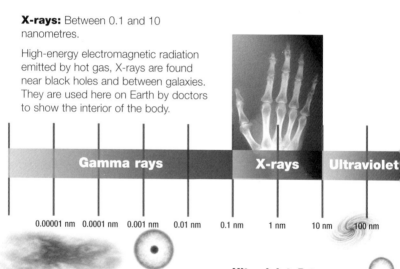

Gamma rays | X-rays | Ultraviolet

0.00001 nm | 0.0001 nm | 0.001 nm | 0.01 nm | 0.1 nm | 1 nm | 10 nm | 100 nm

Gamma rays: Less than 0.1 nanometres.

The most energetic form of radiation, gamma rays are emitted by interstellar gas clouds, neutron star collisions, supernovae.

Ultraviolet: Between 10 and 380 nanometres.

Very hot stars release most of the energy at these wavelengths. The Earth's atmosphere protects us from most of the Sun's harmful ultraviolet radiation.

varying electric and magnetic fields, and are part of a much larger electromagnetic spectrum which extends to longer and shorter wavelengths than visible light. The longer wavelengths are called infrared and radio waves. In principle, they can be indefinitely long, and radio waves with wavelengths of hundreds of kilometres are nothing special. Electromagnetic radiation with wavelengths shorter than visible light is known as ultraviolet light, X-rays and gamma radiation. Any electromagnetic waves shorter than 0.1 billionths of a metre (1×10^{-10} m) is gamma radiation. Gamma rays with wavelengths as short as 10^{-14} m (0.00000000000001 m) are produced in hypernovae (see pages 32–3).

Infrared: Between 750 nanometres and 1 millimetre.

As the name suggests, infrared lies just beyond visible red light. It is emitted by jets of gas from young stars, distant starburst galaxies and cool dust clouds.

Radio waves: Longer than 1 millimetre.

Radio waves are emitted by hot clouds of interstellar gas, hydrogen in the Milky Way and other galaxies, and faint, short wavelengths can be detected from the Big Bang. Radio waves longer than 100 metres are reflected back into space by the ionosphere.

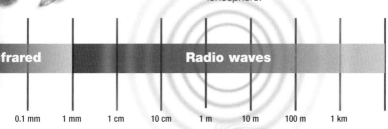

| frared | Radio waves |

| 0.1 mm | 1 mm | 1 cm | 10 cm | 1 m | 10 m | 100 m | 1 km |

Optical: Between 380 and 750 nanometres.

This is light we see, which is made up of seven colours: red, orange, yellow, green, blue, indigo and violet. Their order in the spectrum can be remembered using this phrase: **R**ichard **O**f **Y**ork **G**ave **B**attle **I**n **V**ain.

Small numbers can be big

As we have seen, a number like 7,000,000,000 can be written in scientific shorthand notation as 7×10^9. But it's just as inconvenient to write very small numbers, such as 0.0000000007, as decimals. In a sense, they are also 'big' numbers, because of all the zeros after the decimal point. And there are numbers like 0.1234554321 to worry about. The scientific notation can be adapted to deal with very small numbers by putting a minus sign into the power of 10 (the exponent). Because the power of ten takes account of the decimal point as well as the 0s, 10^{-1} means 0.1, while 10^{-2} means 0.01, and so on. There is one less zero after the decimal point than the number in the exponent. One way of thinking about this is that 10^{-2}, for example, means $1/10^2$, which is 1/100. A number like 0.0000062 can be written as 6.2×10^{-7}, which is the same as $6.2/10^7$ or 6.2/10,000,000. We say 'six point two times ten to the minus seven'. This is very useful when we're talking about things like atoms or viruses. As we have seen, a virus is only a few billionths of a metre across, and we can describe this by saying that its size is a few times 10^{-9} metres.

Atom: **2×10^{-10} m**

Grain of sand: **0.5 x 10⁻³ m**

Grain of pollen: **4 x 10⁻⁷ m**

Virus: **2 x 10⁻⁹ m**

Illustrations not to scale

A mountain of rice

If you put two grains of rice on the first square of a chessboard, four grains on the next square, eight on the third square, and so on, how much rice do you think you would need to put on the last (64th) square? The answer is 2^{64} grains. This is roughly equivalent to the number 18 followed by 18 zeros, or 1.8×10^{19}. (This is vastly more rice than the entire annual production on Earth, which for 2001–2 was estimated to be 589,103,000 tonnes.) To put it another way, this number is about 30 times the age of the Universe since the Big Bang – or 90 times the age of the Solar System – measured in seconds. And it has been achieved with only 64 doublings.

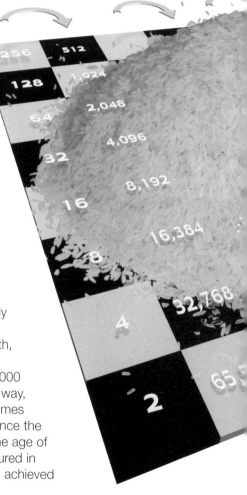

This rapid growth of a number by repeated doubling (or tripling, and so on) is called exponential growth. Exponential growth starts out small (2, 4, 8, and so on) but then turns a corner and shoots upwards dramatically. This doubling at each step means that at each stage of exponential growth you have more of whatever it is that is growing than in all of its past history put together. Human population has grown exponentially, so there are more people alive now than in all previous generations put together.

,072

Infinity

Infinity is a tricky idea to deal with. Think of all the whole numbers (integers) being added up: $1 + 2 + 3 + 4 + 5$ and so on. They add up to infinity. Now imagine doubling every number in the sum and adding them up again: $2 + 4 + 6 + 8 + 10$ and so on. Once again, the answer is infinity. Common sense says that this infinity ought to be twice as big as the first infinity, because every number in the sum is twice as big. But you could say that the second sum is *smaller* than the first one, because the second one leaves out all the odd numbers: 1, 3, 5 and so on! But they are *both* infinity.

These paradoxes are nicely illustrated by the Hotel Hilbert, named for the famous 19th-century German mathematician David Hilbert. Hotel Hilbert has an infinite number of rooms. One night all the rooms are occupied, but when a new guest arrives the receptionist gives her the key to Room 1 after asking the occupant of Room 1 to move to Room 2, the occupant of Room 2 to move to Room 3, and so on. The next night an INFINITE number of new guests arrives, but again the receptionist solves the problem by moving the occupant of Room 1 to Room 2, the occupant of Room 2 to Room 4, the occupant of Room 3 to Room 6, and so on. This frees up an infinity of odd-numbered rooms: 1, 3, 5 and so on.

Zero

The number zero certainly isn't a big number, but it's very important in helping us to write big numbers, and to do calculations involving them. Our counting system uses the same set of numbers over and over again to represent units, tens, hundreds and so on. The place where a digit like 7 appears may mean seven, or seventy, or seven hundred, and so on. Without a symbol for zero to represent an 'empty' place, we couldn't tell the difference between 17, 107 and 1070. The Romans didn't have a symbol for zero, and it's very difficult to do arithmetic using Roman numbers. The earliest mention in history of a symbol for zero comes from a Hindu document written in AD 876, but the idea must be older than that.

Multiplying any number by 0 gives the answer 0, and adding or subtracting 0 from any number gives the number you started with. Dividing any ordinary number by 0 gives the answer infinity, but dividing 0 by 0 is meaningless, and doesn't give you any answer at all – it is *not* equal to 1.

Zeno's paradoxes

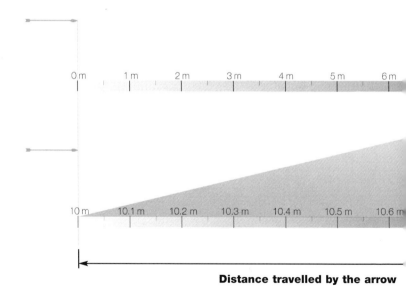

0 m   1 m   2 m   3 m   4 m   5 m   6 m

10 m   10.1 m   10.2 m   10.3 m   10.4 m   10.5 m   10.6 m

Distance travelled by the arrow

Even though zero and infinity are tricky ideas, mathematicians have found ways to use them to solve problems. In Ancient Greek times, thousands of years ago, a philosopher called Zeno posed several puzzles, like this one. If you fire an arrow at a runner 100 metres away, and it's going ten times faster than the runner, by the time the arrow gets there he has moved on (by 10 metres). By the time the arrow covers the extra 10 metres, the runner has moved another metre, and so on. So how does the arrow ever reach him?

Puzzles like this, involving motion, are solved by calculus, developed in the 17th century as a way of adding large numbers of tiny numbers. Using calculus, the flight of the arrow is divided up into an infinite number of infinitely small steps, so that each step seems to be zero size. But the way these tiny steps are added up (called integration) tells you exactly where the arrow is at any moment during its flight, and just when it will catch up with the runner. The same mathematics is used to calculate things like the orbits of the planets, or the volume of a sphere. The answers that come out of the calculations may be quite ordinary numbers, but those answers depend on being able to deal with infinities and zeros in the right way.

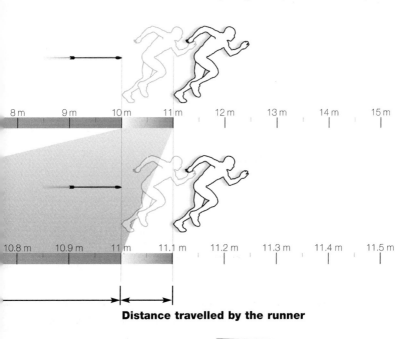

Distance travelled by the runner

More numbers than you can shake a stick at

Whole numbers, like 1, 2, 3, are called integers. But there are also non-integer numbers, like 2/3, in the gaps between integers. A fraction like 2/3 is written as the *ratio* of two integers – one integer divided by another one. So it's called a 'rational' number. But the Ancient Greeks realised that there are some numbers, such as the square root of two (written today as $\sqrt{2}$), which can't be written in this way. Since $\sqrt{2}$ can't be written as a ratio of integers, it's said to be 'irrational'. We can write a number like this as a decimal, but if we do this the number goes on forever without repeating itself – it has an infinite number of digits after the decimal point.

The Ancient Greek philosopher Pythagoras thought that reality was made of mathematics. He and his disciples worshipped numbers, and thought that the world was made up of them. He was finally shocked by his discovery of irrational numbers like pi (see pages 98–9) and $\sqrt{2}$. To Pythagoras, this suggested that the world wasn't mathematically neat and perfect. He even drowned one of his students for revealing this awkward truth to outsiders.

Pi

Pi (or π) is another irrational number. Most people come across pi because it's the ratio of the circumference of a circle (any circle!) to its diameter. But this mysterious number also turns up naturally in many important mathematical equations, including the ones that describe how electricity and magnetism work.

We sometimes use the fraction 22/7 for pi, but this is only an approximation, and the real value of pi is an irrational number with an infinite number of integers after the decimal point. In this sense, all irrational numbers are big numbers. It gets worse. Mathematicians can prove that there is an infinite number of irrational numbers between any pair of rational numbers you can think of. There is an infinite number of irrational numbers between 2/3 and 2/5, an infinite number between 17/256 and 17/257, and so on.

Calculating the circumference of a circle

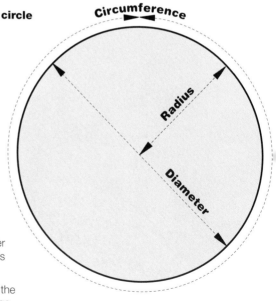

The circumference of a circle is the distance around the outside of the circle. It could be called the perimeter of the circle.

The circumference of a circle can be found by multiplying pi (which is approximately equal to 3.14) by the diameter of the circle.

If a circle has a diameter of 4, its circumference is
3.14 x 4 = 12.56

If you know the radius, the diameter is twice as large.

Pi expressed to 1,000 places

3.14159265358979323846264338327950288419
7169399375105820974944592307816406286208
9986280348253421170679821480865132823066
4709384460955058223172535940812848111745
0284102701938521105559644622948954930381
9644288109756659334461284756482337867831
6527120190914564856692346034861045432664
8213393607260249141273724587006606315588
1748815209209628292540917153643678925903
6001133053054882046652138414695194151160
94330572703657595919530921861173819326117
9310511854807446237996274956735188575272
4891227938183011949129833673362440656643
0860213949463952247371907021798609437027
7053921717629317675238467481846766940513
2000568127145263560827785771342757789609
1736371787214684409012249534301465495853
7105079227968925892354201995611212902196
08640344181598136297747713099605187072113
4999999837297804995105973173281609631859
5024459455346908302642522308253344685035
2619311881710100031378387528865875332083
8142061717766914730359825349042875546873
1159562863882353787593751957781857780532
1712268066130019278766111959092164201989

In with a chance?

The chance of winning the jackpot in the National Lottery in Britain with a single entry is 1 in 14 million. Here's how the maths works. You have to get six numbers right out of 49. If you attempt to guess one number chosen from 49 lottery balls, then the probability that you are correct will be 1 in 49, which can be written as the fraction 1/49. If you have a second attempt, and the previous ball is not replaced, then the probability is 1/48, and so on. So if you choose six numbers in all, then the probability that one of them is the same number as the first ball drawn is 6/49. Given that the first number is chosen correctly, then the probability for choosing the second number correctly is 5/48. The probability of choosing all six numbers correctly is:

6/49 x 5/48 x 4/47 x 3/46 x 2/45 x 1/44

which equals 1/13,983,816

or approximately 1 in 14,000,000

But some people don't realise that because the numbers are selected entirely at random, *any* six numbers between 1 and 49 are just as likely (or unlikely) to win. The birthdays of your six best friends have as much chance as the numbers 1 2 3 4 5 6 and there is just the same chance of the numbers that won last week coming up again this week as any other numbers. You'd be amazed if that ever happened, but it's just as amazing that any particular combination comes up. At least, it would be amazing, if you could guess it first!

So how does anybody ever win? Simple – if the odds are 14 million to 1, and 14 million people enter different sets of numbers, *somebody* is bound to win. If ten horses run in a race, you can be sure that one will win, even if you can't say which one.

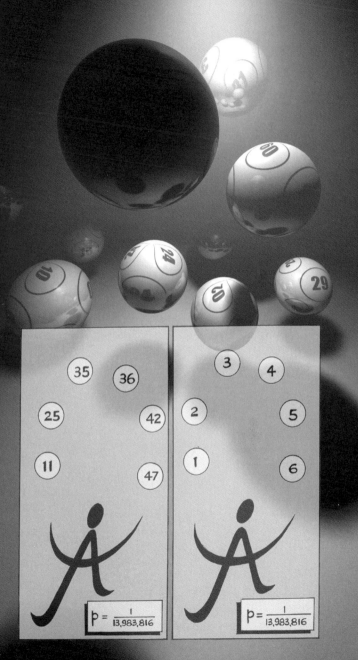

Odds against

Some people worry about flying in aeroplanes, in case they crash. But we all run 'risks' as big as that in everyday life without worrying about it. Everybody has to die sometime, and by comparing the numbers of people who die of different causes, it's possible to work out how dangerous different things are. The chance of being killed as a passenger on a commercial airline flight is 1 in 20,000. This is much better than the chance (if you live in western Europe or North America) of being killed in a car accident (1 in 100) or being accidentally electrocuted (1 in 5,000), but worse than your chance of death by food poisoning (1 in 3,000,000) or (if you live in the USA) of being killed by a tornado (1 in 60,000).

What's the craziest way you might die? Suppose the Earth is hit by a comet or asteroid, like the one that killed the dinosaurs. Astronomers calculate that an impact big enough to kill a quarter of the people on Earth happens once every 375,000 years. Averaging this out, your risk of being killed in this way in any one year is 1 in (4 x 375,000). That's 1 chance in 1.5 million, about ten times bigger than your chance of winning the Lottery. But people live for about 75 years, so your chance of being killed by an impact from space some time in your life is 75 times bigger, and 1.5 million divided by 75 is 20,000. So your chance of being killed by an impact from space is 1 in 20,000 – exactly the same as your chance of being killed in a plane crash. Perhaps flying isn't so dangerous, after all!

Gigabytes and bits

Our everyday counting system is called 'base ten', because it uses ten digits (0, 1, 2, 3, 4, 5, 6, 7, 8, 9). This probably arose because we have ten digits on our two hands. Counting systems can be based on any number of digits. The most important alternative to base ten is

base two, or binary, which is the system used in most computers. In binary, there are just two digits, 0 and 1, which can be represented electronically by a switch being 'off' or 'on'.

In base ten, 1 means 'one', 10 means 'one ten and no digits', 100 means 'one ten squared, no tens and no digits', and so on. In binary, 1 means 'one', 10 means 'one two and no digits', 100 means 'one two squared, no twos and no digits', and so on. In base ten, a 1 with 10 zeros after it means 10^{10}, or 10 billion. In binary, a 1 with 10 zeros after it means 2^{10}, which is 1,024 in base ten. Each

The Earthling is signalling the same number in base ten as the alien in base three.

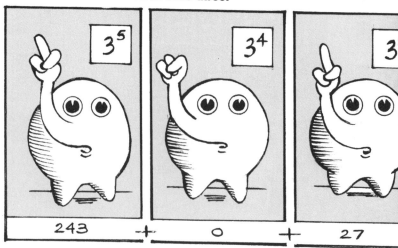

$$90 \quad + \quad 4$$

binary digit (each 1 or 0) is called a 'bit', and eight bits are called a 'byte', rather in the way that ten tens are called a 'hundred'. 2^{10} bytes are actually 1,024 bytes, but this is called a kilobyte (KB) because it is very nearly a thousand bytes. So 2^{20} bytes is a megabyte (MB) and 2^{30} bytes is a gigabyte (GB). If you have a computer with 1 GB of random access memory (RAM), that means there are just under 550 billion (5.5×10^{11}) on/off switches on its memory chips.

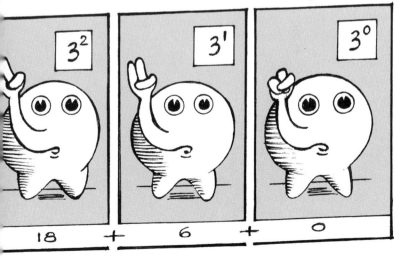

$$18 \quad + \quad 6 \quad + \quad 0$$

429,571,15

446,087,55

06,402,101,809,

Cracking the code: the power of primes

A prime number is any number that can only be divided exactly by itself and 1. So 2, 3, 5 and 7 are primes, but 4, 6, 8 and 9 are not. In 1851, the Russian Pafnuty Chebyshev proved that for any integer (whole number) n there is at least one prime number with a value between n and $2n$. For example, if $n = 6$, $2n = 12$, and in this case there are two prime numbers, 7 and 11, between 6 and 12. Since n can be anything, this means there is an infinite number of primes. Many primes come in pairs, like 11 and 13, or 71 and 73, but some primes occur on their own, with a large gap to the next prime. Overall, primes become rarer and rarer as you look at bigger and bigger numbers. There are 168 primes in the first 1,000 integers, but between the numbers 9,999,001 and 10,000,000 there are only 53 primes. In 1801, the German Carl Gauss proved that any integer bigger than 1 can be broken down into a unique multiple of a set of prime numbers, called factors. So 10 is 2 x 5 (or 5 x 2, the order doesn't matter), 420 is 2 x 2 x 3 x 5 x 7, and so on. This discovery is so important that it's called the fundamental theorem of arithmetic.

446,087,557,183,758,429,571,151,706,402,101,809,886,208,632,412,859,901,111,991,219,963,401,955,724

402,101,809,886,208,632,412,859,901,111,991

429,571,151,706,402,101,809,886,208,632,412,859,901,111,991,219,963,401,955,724

632,412,859,901,111,991,219,963,401,955,724

s easy to find primes using a computer. For example, if you take all e integers bigger than 2 between 1 and 100, you start by throwing way all the numbers bigger than 2 that divide by 2. Now throw way all the numbers bigger than 3 that divide by 3, and so on until ou have thrown out all the numbers that divide by 10. The ones at are left are primes. To make a code that is very hard to crack, ryptographers use very large prime numbers. They find two of ese numbers, using this kind of trick, and keep them secret. Each umber might be 1,000 digits long. Then they multiply the two rimes together to get a huge number, N, 2,000 digits long. This oesn't have to be kept secret, because it would take a computer any years to 'factorise' this huge number. A code message can be urned into numbers by putting A = 1, B = 2 and so on, and the tring of numbers scrambled up using the primes to give a different tring of numbers between 1 and N. The code can be unscrambled nly if you know the two prime numbers.

Freaky fractals

In 1890, the mathematician Giuseppe Peano published a paper which described how to draw a curved line that completely fills a rectangle. This is weird. A line is a 1-dimensional thing – it has length but no width (it is infinitely thin). A rectangle is a 2-dimensional plane surface – it has length *and* breadth. Peano showed how a line could be made to twist and turn inside a plane in such a way that it passed through every mathematical point on the plane (an infinite number of them) without ever crossing itself. A 1-dimensional line seemed to fill a 2-dimensional plane, like a pint filling a quart pot!

Repeat the process

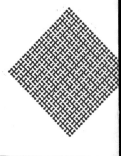

If you take any square in the plane, the Peano curve traces out a set of smaller squares, like tiles, filling the square. And inside each of those squares, it traces out a set of smaller squares still, and so on forever. Each square is exactly like all the others, except for its size. The line is infinitely long, and the pattern of squares within squares goes on forever. It's a 1-dimensional line filling a 2-dimensional plane, and such a thing is now called a 'fractal', meaning it has a fractional dimension, in this case between 1 and 2. The pattern of infinite repetition of identical patterns inside patterns is called self-similarity. There are also fractals which have dimension between 2 and 3 (planes that are crumpled up to fill volumes), and so on.

Scale down by a fact of nine and put nine them together

Divide it into nine smaller squares

Re-arrange them to form a diamond shape

Now cut off the corners indicated

This shape is now the basic template

Re-align them as shown

109

How long is a coastline?

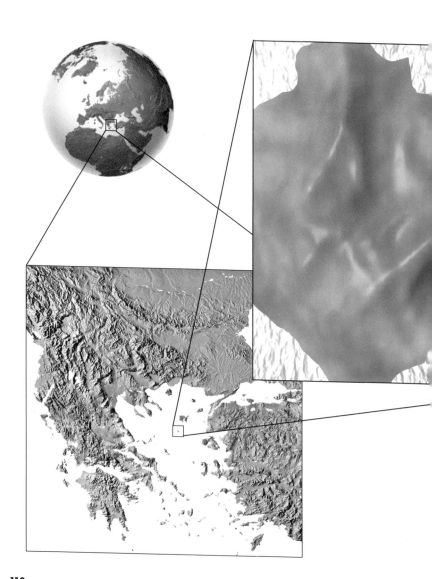

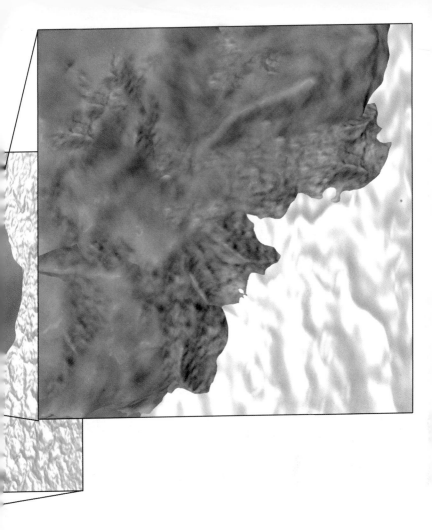

If you drive around a coastline, the milometer in your car will show a certain distance travelled. However, if you walk around the coast path, which follows the line of the cliffs more closely, you will come up with a longer distance. And if you were to climb around the cliff faces and go into every nook and cranny, you would find that the coastline is longer still. In fact, the closer you inspect it, and the more accurate your measurements become, the more the length of the coastline appears to be infinite. The coastline is a fractal formation, and its length is impossible to define exactly.

The butterfly effect

When
scientists talk about chaos,
they don't mean the kind of mess most
people call chaos. They mean that something
is very hard to predict, because what happens next
depends very precisely (they say 'sensitively') on how
things are now. The best example is weather forecasting.
Computers make forecasts by taking account of the
temperature, pressure and so on at different places, and using
these numbers as starting conditions to calculate what the
weather will be like tomorrow. But an American meteorologist
called Edward Lorenz discovered in 1961 that changing one
number in his starting conditions
by a tiny amount could completely change the forecast. He
changed one number in his
computer model from 0.506127 to
0.506, just because he couldn't be
bothered typing in the extra '127'.
That tiny difference made a
huge difference in the pre-
diction. This is sometimes
called the 'Butterfly Effect',
because it tells us that the
weather in London might
be affected by the way a
butterfly flaps its wings in
Florida.

But
the weather is not
always chaotic. Today, fore-
casters always run their computer
models several times with slightly
different starting numbers. If all the
forecasts come out the same, they can
be trusted. But if they are very different,
it means that the weather is chaotic
that week, and the forecasts
are useless.

Written in the sands

One of the biggest puzzles in science is how simple things like atoms (see page 146 onwards) and molecules (two or more atoms joined together) can make complicated things like people and porpoises. This is called complexity, and it depends on huge numbers of simple things working together. A simple way to understand complexity is to look at sand falling steadily onto a table top and piling up one grain at a time. The sand will pile up in a heap, until the slope of the heap reaches a certain steepness. Then, adding more sand will cause a landslide, or a series of landslides, making the sand pile slump down. Eventually, sand covers the entire table and dribbles over the edge when landslides occur. In this state, on average the amount of sand in the pile stays the same, with the same amount falling off the edge of the table as is being added from above. This is called self-organised criticality. All of the complicated patterns in the sand pile are produced by simple grains of sand rubbing past each other – but there are lots of grains of sand.

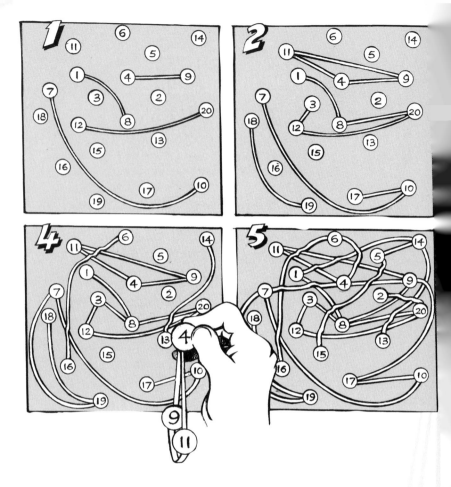

Stuart Kauffman, of the Santa Fe Institute in New Mexico, has another example of what scientists mean by complexity. Imagine a large number of ordinary buttons, perhaps 10,000 of them, spread out across a floor. This is definitely not a complicated system. But look what happens when you start joining the buttons with cotton. Choose a pair of buttons at random and tie them together with a single thread, leaving them in their original positions on the floor. Keep doing this, and if you happen to choose a button that's already connected to another button, don't worry about it – just use the thread to connect it to another button as well. From time to time you'll choose a button that is already connected to another button.

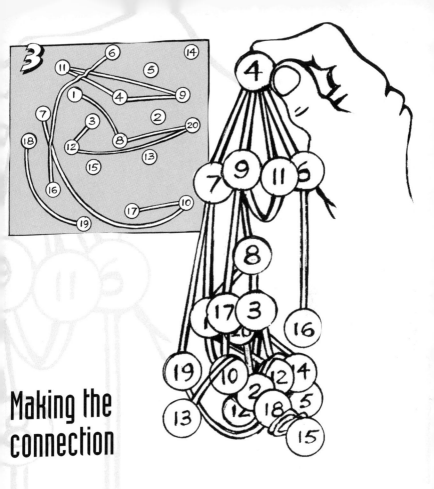

Making the connection

Sometimes, you'll even choose a button that already has two connections, and you'll link it to a fourth one in a growing network of connections.

The way to tell how interesting this network has become is to pick up a few buttons, one at a time, and count the number of connections they each have with other buttons. When the number of threads is at least half the number of buttons, a single supercluster forms, a network in which most of the buttons are linked. The network has stopped changing very much as new connections are made, but it is now, without doubt, a complex system. Tugging one button may affect many buttons far away.

The tangled web

It turns out that life is a bit like a sand pile and a bit like a network. Imagine a frog which eats flies that it catches with a sticky tongue. If some frogs have extra-sticky tongues they will get more food and thrive, having baby frogs with extra-sticky tongues. But if some flies are slipperier than others, they will escape and have extra-slippery babies. So tongues get stickier and flies get slipperier, but you still have the same number of frogs eating the same number of flies. It's like the sand pile and self-organised criticality, where the pile is always changing as grains are added and fall off, but the pile always looks the same.

But there are other things in the web of life. Perhaps fish in the pond also eat flies. If frogs get so good at catching flies that there are none left for the fish, the fish will die out, and perhaps bears that eat fish will go hungry. The stickiness of a frog's tongue might make a bear starve. This web of life is called an ecology. It's like the network of connected buttons. A small change in one part of the ecology can have a big effect somewhere else.

A lot of balls

Here's another example of connections at work. The game of billiards, which is the ancestor of snooker and pool, became popular only in the middle of the 19th century, because until that time people in Europe didn't have a supply of rubber to make the cushions that the balls bounce off at the edges of the tables. The rubber came from the British Empire of the time, and so did another essential ingredient for the game – ivory, to make the balls. The ivory came from the tusks of elephants, and in the middle of the 1850s the game of billiards was so popular that 8,000 elephants were shot in a single year (1854) solely to provide the ivory for billiard balls. People simply didn't realise at the time how bad this kind of destruction was.

Fortunately for the elephants, in the 1860s the first commercial plastic, celluloid, was invented, and began to be used to make billiard balls, among other things. There was a slight problem with celluloid. It was highly inflammable, and if a billiard player hit a very hard shot, there was a chance that the cue ball might explode when it hit another ball. But the elephants didn't seem to mind, and it made the game more interesting.

Our living planet

The atmosphere of Venus is so thick that it contains nearly a hundred times as much gas as the atmosphere of Earth. This gas is mostly carbon dioxide. On Earth, just 0.035% of the atmosphere is carbon dioxide. But there is almost exactly as much carbon dioxide as there is in the air of Venus locked up in carbonate rocks, like limestone, on Earth. It got there because carbon dioxide dissolves in the sea and rainwater, and from there it reacts with other chemicals to form carbonate rocks. Without water, there would be no carbonates. There are no seas on Venus, and no water in its atmosphere, so all of its carbon dioxide stays in the air.

1. Solar radiation passes through Earth's clear atmosphere.

Incoming radiation: 343 watts per m²

2. Some radiation is reflected by Earth's atmosphere and surface.

Outgoing radiation: 103 watts per m²

3. 51% passes through and is absorbed by Earth's surface and warms it: 168 watts per m²

... and some of the Earth's heat is emitted as longwave radiation (infrared) back into the atmosphere.

4. Some of the infrared radiation is absorbed and re-emitted by the greenhouse gas molecules. The direct effect is the warming of the Earth's surface and the lower atmosphere.

Carbon dioxide traps heat from the Sun by the greenhouse effect. When there is more carbon dioxide, the world is warmer. Astronomers know that in the past 3.6 billion years the Sun has got 25% hotter. But geologists know that the temperature stayed more or less the same on Earth for all that time. The reason seems to be that plants control the amount of carbon dioxide in the air. When it gets hotter, plants flourish and take up more carbon dioxide, so the greenhouse effect gets weaker. All life on Earth interacts with its surroundings in this kind of way, as well as with other living things, to make one huge network, a single living planet. This network is known as Gaia.

Gaia

The composition of dry air at sea level on Earth is: nitrogen 78.08%, oxygen 20.95%, argon 0.93%, carbon dioxide 0.03%, neon 0.0018%, helium 0.0005%, krypton 0.0001%, and xenon 0.00001%. In addition to water vapour, air in some localities contains sulphur compounds, hydrogen peroxide, hydrocarbons and dust particles.

Amazingly, the concentration of oxygen is just enough to allow large creatures to exist, but just too low to lead to spontaneous combustion of dry vegetation. And despite being an alarmingly reactive gas, the concentration of oxygen seems to have stayed constant for aeons. Our atmosphere is balanced on an almost-but-not-quite-explosive knife edge, and life seems to like it that way. This has led many people to think that all the living things on Earth somehow act together to influence the whole environment, and that influence helps to maintain the conditions which suit them. This is the thinking behind the concept of Gaia.

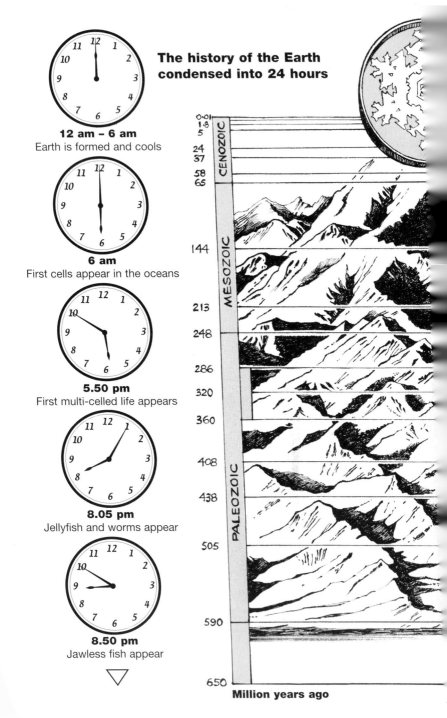

The history of the Earth condensed into 24 hours

12 am – 6 am
Earth is formed and cools

6 am
First cells appear in the oceans

5.50 pm
First multi-celled life appears

8.05 pm
Jellyfish and worms appear

8.50 pm
Jawless fish appear

0·01
1·8
5
24
37
58
65

144

213

248

286
320

360

408

438

505

590

650

CENOZOIC

MESOZOIC

PALEOZOIC

Million years ago

The timescale of life

To put people in perspective, we can imagine shrinking the whole history of the Earth, all 4.5 billion years, into a single day. If we did that, even the dinosaurs wouldn't show up until 11 o'clock at night, and they would be wiped out 20 minutes before midnight. Modern human beings would appear just two seconds before midnight, and all of recorded history (right back to the pyramids) would take place in the last tenth of a second. Changing the analogy, if the lifetime of the Earth were represented by the length of your outstretched arm, all of human history would be wiped out by a single stroke of a nail file across your fingertips. And if the history of the Earth were represented by the height of Mount Everest, your own lifetime would be equal to the thickness of the last snowflake on top of the mountain.

Millions of years ago	Event	Date
4,600	Earth formed	1 January
3,800	Oldest rocks	5 March
3,600	Oldest fossil plants	21 March
2,000	Oxygen in the air	26 July
650	Multi-celled life	10 November
440	First life on land	25 November
230	Dinosaurs evolve	12 December
65	Dinosaurs go extinct	26 December

[All times below are for 31 December]

4	First hominids	07.30
0.1	First *Homo sapiens*	23.49
0.005	Start of recorded history	23.59.34 (26 seconds to midnight)

9 pm
Invertebrates move on land

9.15 pm
Vertebrates begin to move on land

9.17 pm
First amphibians appear

9.37 pm
First reptiles appear

10.27 pm
First dinosaurs appear

Cenozoi

Mesozoi

Paleozo

Precambri

The age of the Earth

Geological time is divided up according to the different kinds of fossils found in the rocks. The layers of rock show the same kinds of fossils for millions of years, then many of these creatures disappear and are replaced by others. This happens when lots of species die out (go extinct) together. The more species that go extinct, the more important the boundary is.

Everything that happened before 590 million years (Myr) ago is one aeon, called the Precambrian. This seven-eighths of Earth history is all lumped together because there are hardly any fossils from it. Everything since then is an aeon called the Phanerozoic. This aeon is divided into three eras. They are called the Paleozoic (from 590 Myr to 248 Myr ago), the Mesozoic (from 248 Myr to 65 Myr ago) and the Cenozoic (from 65 Myr ago to the present). Smaller intervals are used to divide up the eras, like months being used to divide up the year. These are called periods. Periods are divided up into epochs (like months being divided up into days). We live in the Holocene epoch of the Neogene period of the Cenozoic era. To a human being, an epoch is a very long time. But the Holocene began only about 10,000 years ago, and if the whole Phanerozoic aeon was equivalent to a hundred years, the Holocene epoch would be equivalent to just over half a day.

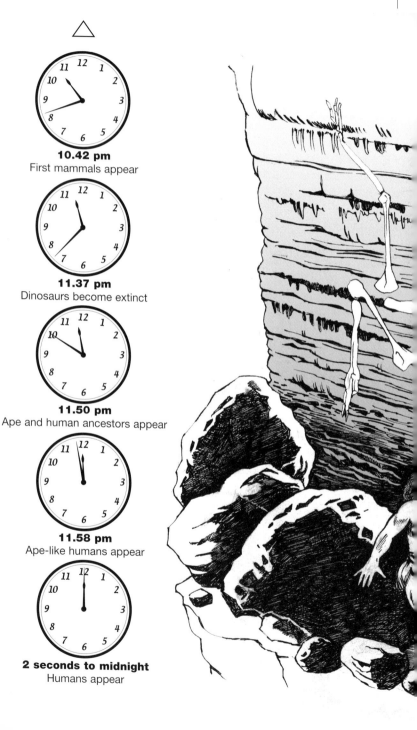

10.42 pm
First mammals appear

11.37 pm
Dinosaurs become extinct

11.50 pm
Ape and human ancestors appear

11.58 pm
Ape-like humans appear

2 seconds to midnight
Humans appear

Dinosaurs without humans

The first dinosaurs evolved from reptilian ancestors about 230 million years ago, at the end of the Triassic period of geological time. Even 230 million years is only 5% of the age of the Earth. Altogether, they ruled the Earth for nearly 150 million years. But that's not much more than 3% of the history of the Earth. Human beings evolved only about 5 million years ago, so we have been around for about 3% of the time that the dinosaurs were around. Contrary to what we see in films like *One Million Years BC*, humans and dinosaurs were not around at the same time.

The age of the dinosaurs was only 3% of the age of the Earth, and the age of humankind (so far) is only 3% of the age of the dinosaurs. That's how old the Earth is, and how young we are.

Dinosaur data

Not all dinosaurs were big, but there are plenty of big numbers associated with them. One of the largest land creatures that ever lived was the Brachiosaurus of the late Jurassic period. It weighed in at up to 100 tonnes and stood about 30 metres long and 20 metres high. Brachiosaurus was a plant-eater, so it didn't need sharp teeth; but carnivores like Allosaurus, which lived around the same time (and probably ate Brachiosaurus), had teeth more than 15 centimetres long with sharp, serrated edges like a saw. It could tear off chunks of flesh from its prey in one gulp.

The fastest dinosaur, Ornithomimus, was built like an ostrich. About 3.5 metres long, it could run at speeds up to 80 km per hour in short sprints. Some dinosaurs had heads with very thick skulls that they probably used like battering rams in fights. Pachycephalosaurus (the name means 'thick-headed lizard') had a skull 25 cm thick. Some dinosaurs were immensely long. From the tip of its head to the end of its tail, Diplodocus stretched for nearly 30 metres, although it weighed only about 12 tonnes. It was longer than three buses in a row. But dinosaurs were not very bright. The brain of Apatosaurus, a 20-tonne vegetarian, weighed only 0.001% as much as its body.

Blasts from the past

The age of the dinosaurs was brought to an end by the impact of a rock from space with the Earth 65 million years ago (see pages 40–1). But this wasn't the first blast from space to cause extinctions of life on Earth, and it won't be the last. By counting the number of craters on the surface of the Moon, our near neighbour, astronomers calculate that there have been at least 40 impacts as big as the one that killed the dinosaurs since life began on Earth 4 billion years ago.

But even smaller impacts could be bad news for us. In 1908, a piece of a comet, streaking into our atmosphere and overheating, exploded in the air above Tunguska, in Siberia. It flattened the forest over an area of more than 2,000 square kilometres, with a blast estimated as equivalent to 20 megatonnes of TNT, a thousand times as powerful as the nuclear bomb that destroyed Hiroshima in 1945. The shock produced vibrations in the crust of the Earth recorded 4,000 km away in St Petersburg. It was caused by a lump of rock or ice weighing 100,000 tonnes hitting the atmosphere at a speed of 30 km per second (more than 100,000 km per hour).

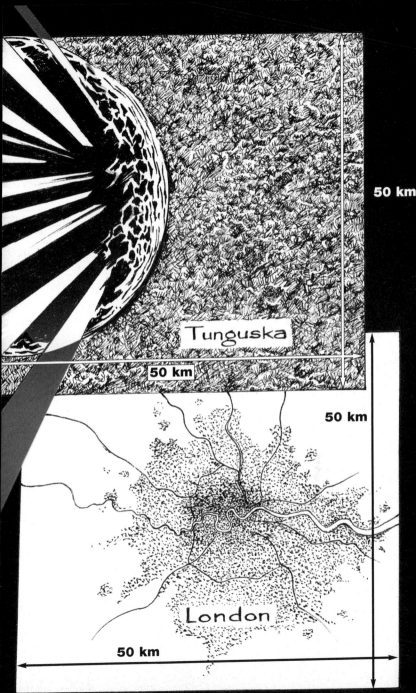

50 km

Tunguska

50 km

50 km

London

50 km

The thin green smear

Mount Everest in the Pacific Mariana Trench

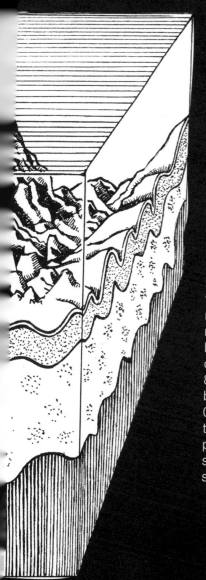

Rocks from space are bad news for life on Earth because the 'life zone' of our planet is so thin. The rock that killed the dinosaurs was only about 10 km across. But the average depth of the sea is only 3.7 km. A rock 10 km across could sit in the sea with more than half of it above the waves. Even the floors of the deepest trenches in the ocean bed are only 11 km deep. (The North Atlantic is just over 3 km deep, but it's 4,800 km wide – to have the same proportions, a puddle 3 millimetres deep would have to be nearly 5 metres across.) The peak of the highest mountain on Earth, Mount Everest, is nearly 9 km above sea level. So, from the bottom of the deepest ocean trench to the top of the tallest mountain is just 20 km. This is the region in which life exists. If the Earth were shrunk from its diameter of just under 13,000 km to a ball 80 cm across (the size of a beach ball), the life zone would be only 0.12 mm thick – about the same as the thickness of the paper on this page. This is why the life zone is sometimes called a 'thin green smear' on the surface of our planet.

The oceanic conveyor belt

The Gulf Stream is like a river of warm water flowing on the surface of the ocean up the western side of the North Atlantic, carrying thirty million cubic metres of water every second. It's part of a global system of ocean currents which flows all the way from the tropical Pacific around South Africa's Cape of Good Hope, picking up warmth from the Sun for most of its long journey. Because warm water is less dense than the cold water of the deeper ocean, it forms a surface current. But the water gets more and more salty, because evaporation carries water away into the air. In the far north of the

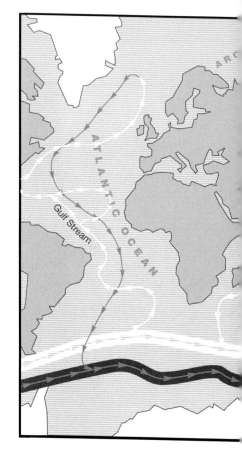

Atlantic, the current gives up its heat to the winds which blow from west to east at those latitudes, carrying the warmth towards Europe. But the current gets colder and more dense. With the extra saltiness as well, this makes it sink into the depths, where it returns all the way to its starting point before welling up again in the North Pacific. The whole system forms a kind of conveyor belt, driven by upside-down convection, pushed by the descending dense, salty water of the North Atlantic. The flow of this 'river' in the ocean is twenty times greater than the flow of all the rivers on all the continents of the Earth put together.

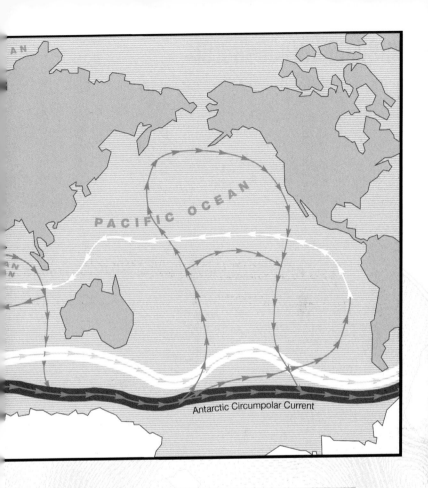

PACIFIC OCEAN

Antarctic Circumpolar Current

Cold, salty, deep current

Warm, less salty, shallow current

Keeping Europe warm

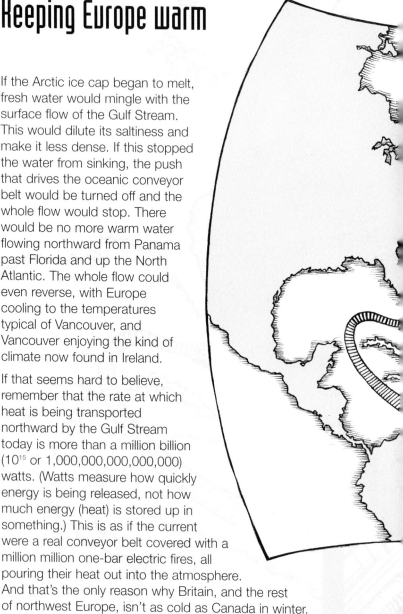

If the Arctic ice cap began to melt, fresh water would mingle with the surface flow of the Gulf Stream. This would dilute its saltiness and make it less dense. If this stopped the water from sinking, the push that drives the oceanic conveyor belt would be turned off and the whole flow would stop. There would be no more warm water flowing northward from Panama past Florida and up the North Atlantic. The whole flow could even reverse, with Europe cooling to the temperatures typical of Vancouver, and Vancouver enjoying the kind of climate now found in Ireland.

If that seems hard to believe, remember that the rate at which heat is being transported northward by the Gulf Stream today is more than a million billion (10^{15} or 1,000,000,000,000,000) watts. (Watts measure how quickly energy is being released, not how much energy (heat) is stored up in something.) This is as if the current were a real conveyor belt covered with a million million one-bar electric fires, all pouring their heat out into the atmosphere. And that's the only reason why Britain, and the rest of northwest Europe, isn't as cold as Canada in winter.

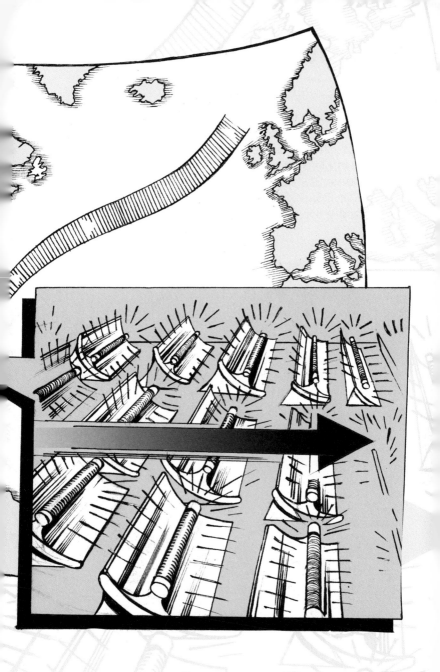

The big freeze

The permanent ice caps of the Earth cover 10 million square kilometres in the north and 14 million square kilometres in the south. But because there is land around the northern ice caps where snow can settle in winter, at that time of year the northern area of ice cover reaches 50 million square kilometres. The southern equivalent reaches only 20 million square kilometres, because there is very little 'spare' land in the deep south.

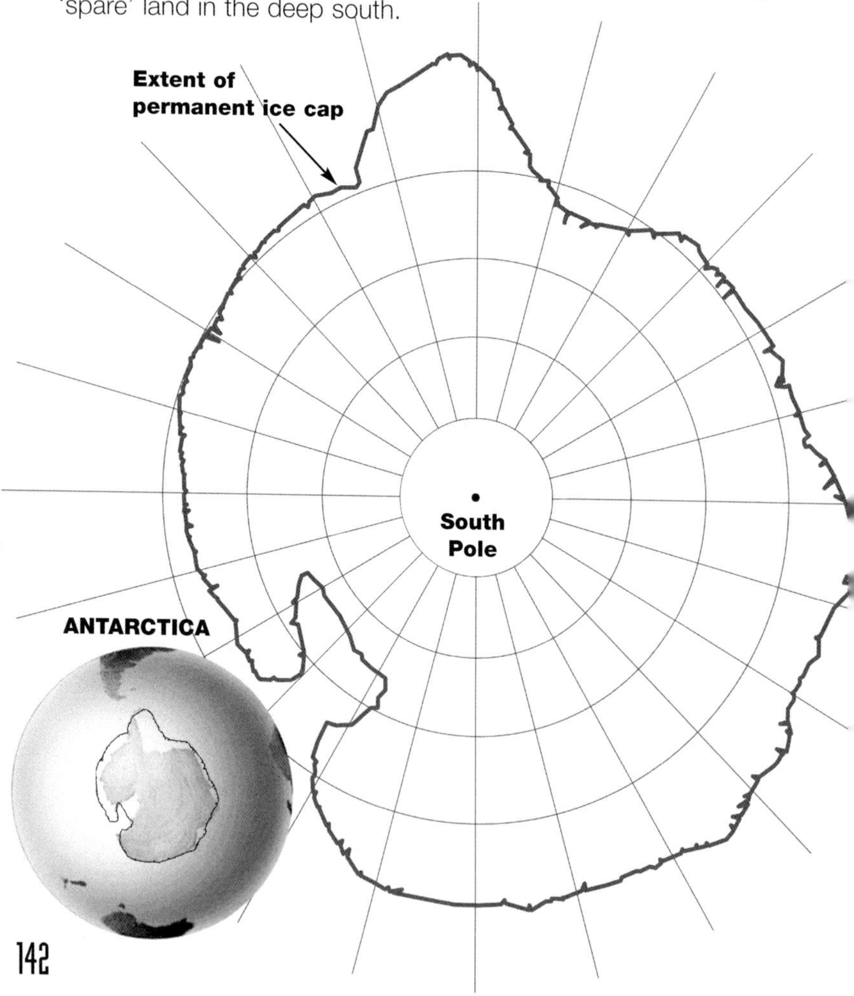

Extent of permanent ice cap

South Pole

ANTARCTICA

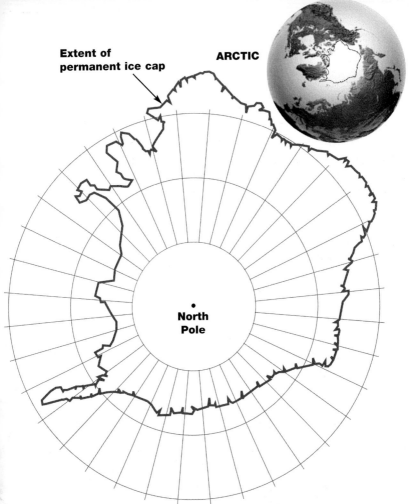

Extent of
permanent ice cap

ARCTIC

North
Pole

During the Ice Age which ended about 15,000 years ago, all of the region of the Northern Hemisphere where snow falls regularly today was covered in ice all year round. The ice sheets covered 50 million square kilometres. Summer in the Ice Age was like winter is today. In America, the biggest ice sheet stretched from the Rocky Mountains to Newfoundland, and as far south as southern Illinois. In Europe, ice covered most of Great Britain and stretched across Germany and Poland. Further east, there was no thick ice sheet because the air was so dry that there was very little snowfall. But to the south there were large glaciers covering mountains such as the Alps.

Making icebergs

It takes a lot of snow to make an iceberg. An iceberg is born when a slow-moving river of ice, called a glacier, reaches the sea and pieces break off it. In the region of Greenland where the glaciers form, between 50 and 150 cm of snow fall each year. But as the snow gets squashed into ice by the weight of more snow falling on top of it each year, each centimetre of ice corresponds to many centimetres of snow. The oldest ice in Greenland, at the bottom of the deepest ice sheets touching the bare rock, fell as snow about 150,000 years ago, during the Ice Age before last, and is buried nearly 2 km below the surface of the ice.

Icebergs are not that old, because they come from the edges of the ice sheet where the ice is thinner. The average age of the ice in a 'berg when it breaks off from the glacier (calves) is about 5,000 years. Most of the icebergs calved from Greenland melt within two years, because the waters of the North Atlantic are warmed by the Gulf Stream. But in the colder waters around Antarctica, giant icebergs may take ten years to melt away.

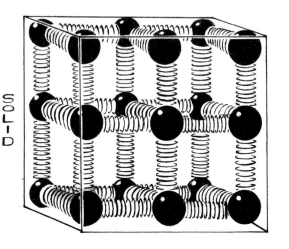

S O L I D

L I Q U I D

Hot and cold

Temperature is really a measure of the movement of atoms and molecules in a substance. Heat gives atoms energy. In a solid, atoms are held in place by bonds between them, but they still manage to vibrate as if attached to each other by springs. When this vibration gets to a certain point, the bonds between the atoms are no longer able to hold onto them, and the atoms begin to move more easily. When this happens, the solid becomes a liquid. The atoms are still constrained a little, but as the substance continues to get hotter, the atoms whizz around faster and eventually are able to break free of all their constraints. When this happens, the substance changes from a liquid to a gas.

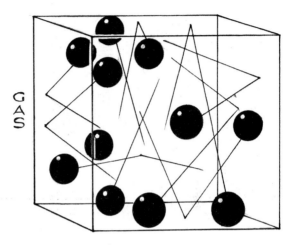

So, the lower a temperature, the less atoms move about. At –273°C they are moving as slowly as it is possible to move, a kind of running on the spot. This is the coldest temperature possible. Because this is the so-called Absolute Zero of temperature, scientists often use a temperature scale called the Kelvin scale (K), which starts at the coldest point possible, absolute zero (–273°C), and has the freezing point of water at 273 K and the boiling point of water at 373 K.

Scientists have been able to cool liquid helium to temperatures less than a billionth of a degree (0.000000001 K) above absolute zero. This is very difficult, and can be done only in specialist laboratories, but it's relatively easy to cool liquid helium to about 4 K (–269°C). At the other extreme, in experimental reactors known as Tokamaks, scientists have been able to achieve temperatures roughly the same as those at the heart of the Sun, about 15 million K, for brief moments. These limits are the extremes of temperature that occur on Earth today.

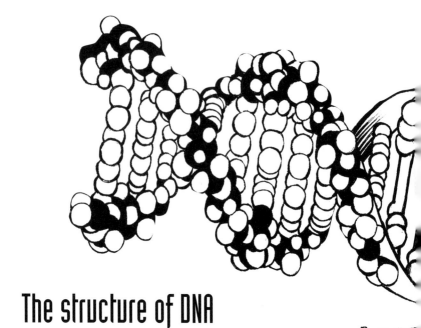

The structure of DNA

Life depends on the complexity that occurs
when lots of very simple things interact with
each other in a network (see pages 114–15).
The simple things are atoms, and the interacting
networks of atoms are called molecules. The
most important life molecule is DNA. Each
molecule of DNA is made of two long strands,
strings of atoms, joined together by cross-links
like the rungs of a ladder. Each cross-link (rung) is
called a base pair. Inside a cell, the DNA is packed
into little rod-shaped structures called chromosomes.
People have 46 chromosomes in each cell. Inside the
chromosomes, different stretches of DNA correspond to
different genes, each of which plays a special part in controlling
the chemistry of life. All of the DNA in a human cell (called its
genome) contains about 30,000 genes, so there must be, on
average, more than a thousand genes on each chromosome.

In the whole human genome, there are three billion base pairs linking strands of DNA – three billion rungs on the DNA ladder. There are four different kinds of rungs, and these work like letters in a four-letter code, or alphabet, to store information. So the DNA 'book' with instructions on how to run your body contains three billion letters. For comparison, this book contains a bit more than 100,000 letters. So the book of life is 30,000 times as big as this book.

The size of DNA

It isn't obvious how much information is stored in the book of life, because atoms and molecules are so small. Each strand of DNA is only one tenth of a billionth of a centimetre (0.0000000001 cm) wide, but if all the DNA in a single human cell were uncoiled and laid end to end to make a single strand, it would stretch for 175 cm. If you are less than 175 cm tall, the DNA from a single cell in your body could be stretched out to more than your own height. Looking at this another way, if the thickness of the strands of DNA in a single cell could be increased to the thickness of an average violin string, and the length increased in proportion, the strand would be more than 15 kilometres long.

Remember, though, that there are a hundred thousand billion cells in the human body (see pages 56–7). And each of those cells contains 175 cm of DNA! If you could take all the DNA in your body and stretch it out in this way, it would reach for a truly astronomical distance. It would stretch for 175 billion kilometres, enough to reach from here to the planet Pluto (the most distant planet in the Solar System) three times over.

Send in the clones

Thanks to genetic engineering, the word 'clone' is often used these days to refer to a single individual that is an exact copy of another individual. When the genetic material, the DNA, of a sheep was used to make a baby sheep, the lamb (called Dolly) was said to be a clone of her mother.

But all the excitement about Dolly was because it was an *animal* that was being cloned. Farmers and horticulturists have known about cloning for a long time, because it's easy to make copies of plants. If you take a cutting from a plant, and use the cutting to grow a new plant, it will have exactly the same DNA as its 'parent', and that's what cloning is all about.

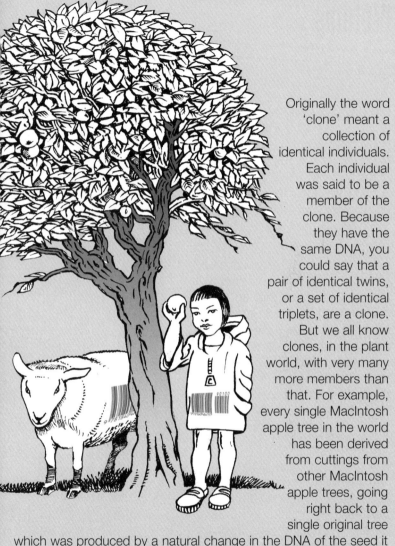

Originally the word 'clone' meant a collection of identical individuals. Each individual was said to be a member of the clone. Because they have the same DNA, you could say that a pair of identical twins, or a set of identical triplets, are a clone. But we all know clones, in the plant world, with very many more members than that. For example, every single MacIntosh apple tree in the world has been derived from cuttings from other MacIntosh apple trees, going right back to a single original tree which was produced by a natural change in the DNA of the seed it grew from (a mutation). Although the original MacIntosh apple tree is long dead, all of its descendants carry exact copies of that DNA, and they are all members of a single clone. Something to think about next time you bite into an apple.

Proteins

DNA in chromosomes is like a book which stores all the information about how to run your body. But the molecules that actually do the work of the cells are called proteins. The coded message of one gene in a chromosome tells the cell how to make one protein.

Each protein is made up of units called amino acids. A typical protein molecule contains two or three hundred amino acids joined together in a chain. Some are much smaller, some much bigger. One kind of protein molecule, called titin, found in muscle, contains 27,000 amino acids linked together in a single chain. These chains can fold up on themselves to make compact balls (like the toy known as Rubik's snake).

A good example of a protein at work in the body is haemoglobin. This is a protein made of about 140 amino acid units, which folds into a ball and is carried in the red cells of your blood. It picks up oxygen from the lungs and carries it around in the blood to wherever it's needed.

A haemoglobin cluster showing a haem group holding a single iron atom. At the binding site on the iron atom is the oxygen atom.

Oxygen atom

Haemoglobin cluster

Iron atom

Haem group

Counting atoms

One person who was puzzled by the way that things of the same size can have different weights was Edmund Halley, who gave his name to a comet. In the 17th century, he realised that this meant atoms are very small, with lots of space between them. Gold is seven times as dense as glass, so Halley said there must be at least six-sevenths 'empty space' between the atoms in glass.

When goldsmiths coat a silver wire with gold, they pull the wire through the gold to make a layer of it on the surface. This layer must be at least one atom thick, or the silver would show through. Because he knew how much gold was used, and the diameter and length of the piece of wire, Halley calculated that a cube of gold with sides one-hundredth of an inch long must contain at least 2,433 million atoms. This estimate is actually much too small (Halley realised this), but the amazing thing is that it was published in 1601, more than four hundred years ago.

Edmund Halley

Weighing molecules

In the 19th century, two hundred years after Halley, an Italian called Amadeo Avogadro found a better way to count atoms. In a box of gas, the pressure on the walls is caused by atoms hitting the walls and bouncing off. If the temperature is the same, and the pressure is the same, the number of atoms or molecules in the box must be the same, whatever the gas is made of (remember that a molecule is just a collection of atoms joined together in a certain way). The only difference is that some things weigh more than others. Molecular weights are measured on a scale where one hydrogen atom weighs one unit. Each oxygen atom weighs 16 on this scale, and there are two oxygen atoms in each molecule (that's why it's written as O_2), so the molecular weight of oxygen is 32. It turns out that if you take an amount of any substance equal to this atomic weight in grams (32 grams of oxygen, for example), it will contain the same number of molecules – 6×10^{23} (600,000,000,000,000,000,000,000) molecules. This huge number is now known as the Avogadro Constant. Because each oxygen molecule is made up of two atoms, 32 grams of oxygen will actually contain 12×10^{23} atoms, and so on.

Period

		1 IA 1A	
1		**H** 1.008	2 IIA 2A
2		**Li** 6.941	**Be** 9.012
3		**Na** 22.99	**Mg** 24.31
4		**K** 39.10	**Ca** 40.08
5		**Rb** 85.47	**Sr** 87.62
6		**Cs** 132.9	**Ba** 137.3
7		**Fr** (223)	**Ra** (226)

ic Table of the Elements

					18 VIIIA 8A

13 IIIA 3A	14 IVA 4A	15 VA 5A	16 VIA 6A	17 VIIA 7A	2 He 4.003
5 B 10.81	6 C 12.01	7 N 14.01	8 O 16.00	9 F 19.00	10 Ne 20.18

6 VIB 6B	7 VIIB 7B	8 ------- VIII ------- ------- 8 -------	9	10	11 IB 1B	12 IIB 2B	13 Al 26.98	14 Si 28.09	15 P 30.97	16 S 32.07	17 Cl 35.45	18 Ar 39.95
24 Cr 52.00	25 Mn 54.94	26 Fe 55.85	27 Co 58.47	28 Ni 58.69	29 Cu	30 Zn	31 Ga	32 Ge	33 As 74.92	34 Se 78.96	35 Br 79.90	36 Kr 83.80
42 Mo 95.94	43 Tc (98)	44 Ru 101.1	45 Rh	46 Pd					51 Sb	52 Te 127.6	53 I 126.9	54 Xe 131.3
74 W 183.9	75 Re										85 At	86 Rn (222)

56g iron

207g lead

**Amadeo
Avogadro**

Measuring molecules

Modern scientists have several ways to measure the sizes of atoms or molecules in different kinds of stuff. One way is to compare liquids and gases. In a liquid, all the molecules are touching one another. In a gas, they fly about and bounce off one another, with lots of empty space in between. Experiments with gases tell you how big the Avogadro Constant is, then measuring the volume of liquid and dividing by the appropriate number tells you how big each molecule is. Air is a mixture of molecules of different kinds. The first experiments along these lines, in the 19th century, showed that a typical molecule of air is a few millionths of a millimetre across – about 0.000000005 of a metre. Modern experiments confirm this, and show that all atoms are roughly the same size. If we laid out each atom in 82 grams of lead (82 being the atomic number of lead), side by side so that they were touching each other, they would stretch from the Earth to the edge of the Solar System and back nearly ten times.

The speed of air

Air is mostly empty space, but it seems to exert a uniform pressure on us. Anyone who has ever tried to pump up the tyre of a bicycle knows how much pressure it takes to fill the tyre with air. The reason is that although the molecules of air are tiny, there are billions of them moving very fast, so they are always bouncing off your skin, or the walls of the tyre, to make a steady pressure. To do this, they have to move fast and collide with one another very often. Atoms move faster at higher temperatures, but at 0°C an oxygen molecule in air is moving at 461 metres per second. That's 1,660 km per hour, or roughly a thousand miles per hour. But it doesn't get a chance to go far in a straight line. On average, for the density of air at sea level and 0°C, each molecule travels only 13 millionths of a metre between collisions. So each molecule of oxygen in the air on a freezing day by the seaside is involved in more than 3.5 billion (3,500,000,000) collisions every second. If the temperature is higher, the speed goes up and there are even more collisions each second.

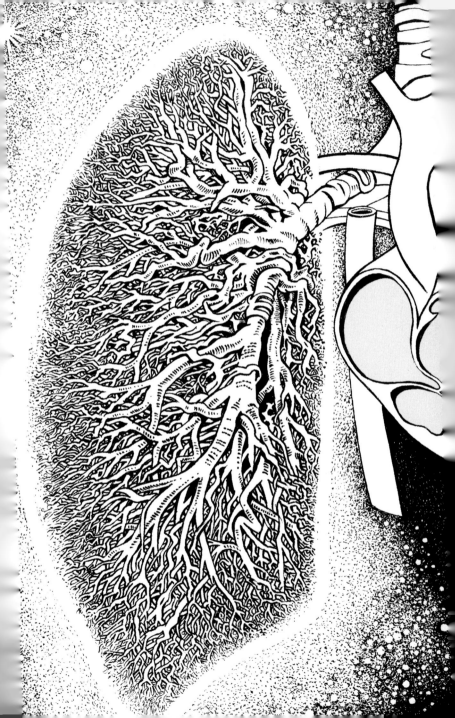

Every breath you take

The Avogadro Constant is a good example of why people refer to big numbers as 'astronomical'. A number like 6×10^{23} doesn't really mean anything at all to most people. But there are that many molecules of gas in just 32 grams of oxygen. Under the standard conditions of temperature and pressure, that much oxygen takes up just thirteen litres in volume. The total number of stars in all the galaxies like the Milky Way in the entire visible Universe is 'only' 10^{23}. So a couple of litres of oxygen at standard temperature and pressure contains as many molecules as there are stars in all the galaxies visible to our best telescopes put together. Air is mostly made of nitrogen and oxygen. The difference in size of nitrogen and oxygen molecules is not enough to worry about for this rough calculation, so a couple of litres of the air that you breathe contains as many molecules as there are stars in all the galaxies visible to our best telescopes put together. The maximum capacity of the lungs of an adult person is about six litres; but even a child who takes a deep breath has more molecules of air in their lungs than there are stars in the visible Universe. That's about 50,000,000,000,000 molecules for *each* person alive on Earth today.

All the molecules get mixed up in the air when they are breathed out, and eventually spread right through the atmosphere. This means that every time you take a breath, it includes a few molecules breathed out by everyone who has ever lived, including Julius Caesar, Marilyn Monroe, Genghis Khan and Eminem.

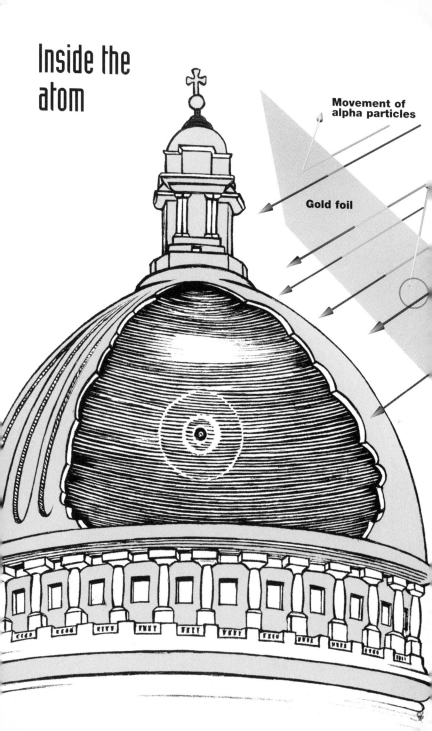

Inside the atom

Movement of alpha particles

Gold foil

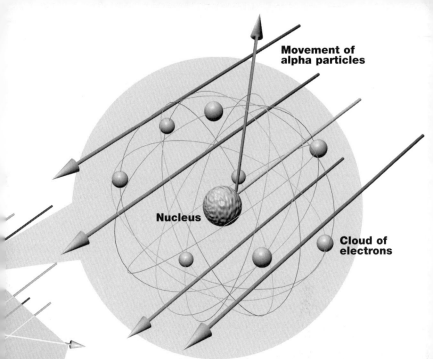

Movement of alpha particles

Nucleus

Cloud of electrons

If you think atoms are small, you ain't seen nothin' yet. At the beginning of the 20th century, scientists in Manchester probed atoms by shooting beams of tiny particles (called alpha particles, or alpha rays) at thin sheets of gold foil. Most of the alpha particles went straight through the gold foil as if it weren't there; just a few bounced back when they hit something solid inside the gold. From these experiments, the physicist Ernest Rutherford worked out that atoms are mostly empty space. There is a tiny central nucleus (the bit that the alpha particles bounce off) surrounded by a cloud of electrons (see page 168), rather like the way the planets orbit the Sun. The alpha particles brush through the electrons as if they weren't there. The experiments show that although an atom is about 10^{-8} cm across, the nucleus is only 10^{-13} cm across. The difference looks small, but remember that it's 10^5, or 100,000. The diameter of the nucleus is just one-hundred-thousandth of the diameter of the whole atom. This is equivalent to the size of a pinhead compared with the size of the dome of St Paul's Cathedral in London.

Inside the nucleus

The particles inside atoms are so ridiculously small that it makes more sense to talk about them in terms of their mass (weight) rather than their diameter. The electron is the smallest of them all. It is a truly fundamental particle and cannot be divided up into anything else. It has a mass of just 9×10^{-31} kilos (0.00000000 00000000000000000000009 kg), and is negatively charged. Because mass is equivalent to energy, according to Einstein's famous equation $E = mc^2$, physicists often measure these tiny masses in electron Volts. The mass of the electron is 0.5 million electron Volts (0.5 MeV).

Inside the nucleus, there are just two kinds of particles. The proton has positive electric charge and weighs in at 1.7×10^{-27} kilos, or 938 MeV, making it roughly two thousand times as massive as an electron. The other kind of nuclear particle, the neutron, has no electric charge and a mass of 940 MeV, just a tiny bit more than the proton's mass. But protons and neutrons are not truly fundamental particles; they are each made up of even smaller particles, called quarks, which are thought to be fundamental. There are three quarks in each neutron, and three in each proton.

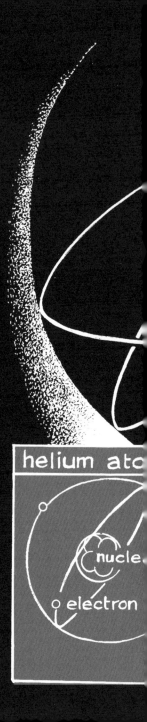

helium ato

nucle.

o electron

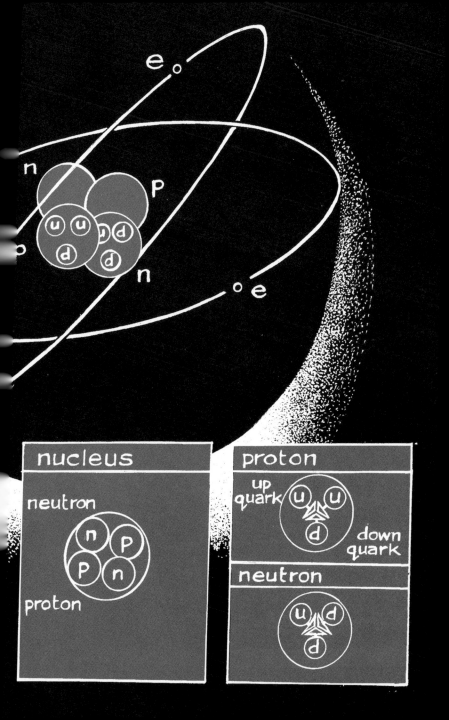

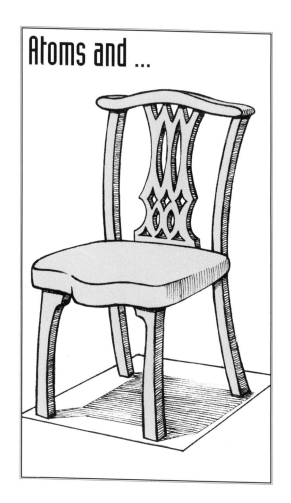

Atoms and ...

There's even less to atoms and nuclei if you think in terms of volume. Atoms and nuclei are more or less round, and the volume of a sphere is proportional to the cube of its radius. Because the radius of the nucleus is 10^{-5} times the radius of an atom, this means that the volume of the nucleus is only 10^{-15} times the volume of the whole atom. In everyday language, this means that only one millionth of a billionth of an atom is solid nucleus.

... empty space

In a solid object, like the chair you are sitting on, the atoms are joined to one another because the electron clouds link up, as if the atoms were holding hands. But inside the atoms there is empty space all the way down from the electrons to the nucleus. Since everything on Earth is made of atoms, that means that not only your chair but your own body and everything else you come into contact with is made up of a million billion times more empty space than solid matter. The only reason things seem solid is that they are held together by electric forces that operate between the various particles involved.

Probing atoms

People know what goes on inside atoms and molecules because they have been probed using beams of particles accelerated to very high speeds. These particles must be electrically charged, so that they can be gripped in electric and magnetic fields which are used to speed them up. Sometimes, electrons themselves are used. The particles are boosted to close to the speed of light, which means they are carrying an enormous amount of energy. Then, they are directed to hit a target, or made to collide with another beam of particles going the other way. The debris from the collision sprays out in all directions, and is picked up by different kinds of detectors. Then scientists try to work out what the targets were made of before the collision. It's a bit like trying to work out how a car is built by looking at the wreckage from a motorway pile-up.

Accelerating particles

There are two main kinds of particle accelerator. Linear accelerators accelerate particles in straight lines. The Stanford Linear Accelerator (SLAC), in California, has a tube 3 km long (which actually goes right under a freeway) aligned straight to an accuracy of 0.5 mm. It has all the air pumped out to make a vacuum, and beams of electrons are accelerated down it by electric fields (like surfing the radio waves) to energies of 50 billion electron Volts each.

The other kind of accelerator whirls particles round and round in a circle until they have as much energy as the experimenters need (or as much as the machine can provide). The most powerful accelerator is the Large Hadron Collider (LHC), at the CERN research institute in Geneva. It's built in a tunnel 27 km in diameter dug under the Swiss Alps, and is due to begin operating at energies as high as 14,000,000,000,000 eV in the middle of the first decade of the 21st century. With the LHC running, CERN uses almost half as much power as the entire city of Geneva. This is pumped into protons going round and round the ring until the handful of protons has as much energy as the US Navy's biggest aircraft carrier travelling at a speed of 11 knots. The energy in the colliding beams is like two aircraft carriers meeting head on.

Collision within a particle accelerator

Route taken by a particle accelerator at Westerhorn near Hamburg in Germany

Radioactive decay

Some kinds of atomic nuclei are unstable. They spit out particles and turn themselves into different kinds of nuclei. This is called radioactive decay. Radioactive substances decay in a curious way. No matter how much (or how little) you start with, exactly half the nuclei decay in a certain time, which is different for each radioactive substance. This is called the half-life. If you start with 256 trillion atoms of the same radioactive stuff, in one half-life 128 trillion will decay, in the next half-life 64 trillion will decay (half of what was left), in the third half-life 32 trillion will decay, and so on.

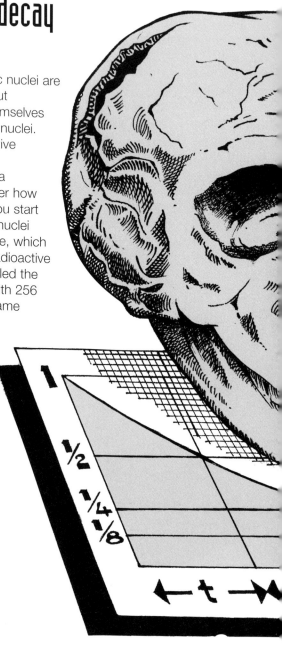

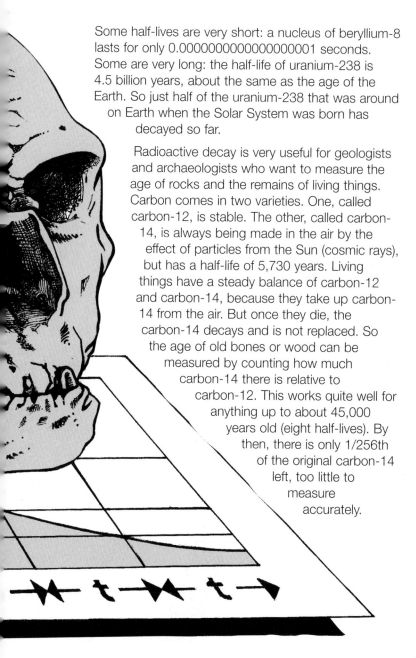

Some half-lives are very short: a nucleus of beryllium-8 lasts for only 0.0000000000000000001 seconds. Some are very long: the half-life of uranium-238 is 4.5 billion years, about the same as the age of the Earth. So just half of the uranium-238 that was around on Earth when the Solar System was born has decayed so far.

Radioactive decay is very useful for geologists and archaeologists who want to measure the age of rocks and the remains of living things. Carbon comes in two varieties. One, called carbon-12, is stable. The other, called carbon-14, is always being made in the air by the effect of particles from the Sun (cosmic rays), but has a half-life of 5,730 years. Living things have a steady balance of carbon-12 and carbon-14, because they take up carbon-14 from the air. But once they die, the carbon-14 decays and is not replaced. So the age of old bones or wood can be measured by counting how much carbon-14 there is relative to carbon-12. This works quite well for anything up to about 45,000 years old (eight half-lives). By then, there is only 1/256th of the original carbon-14 left, too little to measure accurately.

The long and winding wall

There's a popular myth that the Great Wall of China is the only man-made object visible from space, because it's so long. But this is just a myth. What matters when it comes to seeing the wall from above isn't its length but its width – and this is less than 6 metres, nowhere near big enough to see from space. After all, a human hair a metre long isn't any easier to see across a room than one a centimetre long!

But the Great Wall really *is* long. Construction started in the 7th century BC and carried on for hundreds of years, with as many as

600,000 labourers working on the project at the height of activity in the 3rd and 2nd centuries BC. Then, between about 1368 AD (the start of the Ming dynasty) and 1550 AD, the wall was completely renovated into the state we see today. Until 2001, its length was officially given as 6,500 kilometres. But then a team of Chinese archaeologists uncovered the remains of an extension of the wall into the Lop Nur region, which is now a desert but was once fertile. So the total length of the wall is now known to be 7,200 km.

By the way – astronauts really have looked to see if they could indeed pick out the wall from space. We met one and asked him. They couldn't.

Some Big Events

15 billion years ago: the Universe is born in the Big Bang

10 billion years ago: light from the most distant galaxies that we can see using the Hubble Space Telescope begins its journey towards Earth

5 billion years ago: the Sun comes into existence

4.6 billion years ago: the Earth forms

3.8 billion years ago: the oldest rocks form

3.6 billion years ago: plant life begins

2 billion years ago: oxygen becomes present in the Earth's atmosphere

650 million years ago: multi-celled life evolves

440 million years ago: the first life appears on land

230 million years ago: the first dinosaurs evolve

65 million years ago: the planet plunges into an ice age, and the dinosaurs become extinct

10 million years ago: the photons now being emitted by the Sun begin their journey from the heart of the Sun to its surface

4 million years ago: the first hominids (the primate ancestors of humans) evolve

2.4 million years ago: oldest known stone tools

100,000 years ago: humans (*Homo sapiens*) evolve

53,000 years ago: colonisation of Australia

40,000 years ago: *Homo sapiens* inhabit Europe

15,000 years ago: colonisation of North America begins

6500 BC: farming begins in Europe

4500 BC: copper-working begins in the Balkans

4000 BC: horses domesticated in the Ukraine

1650 BC: bronze ploughs in use in Vietnam

1200 BC: iron-working developed

7th century BC: construction of the Great Wall of China starts

508 BC: Pythagoreans teach that the Earth is a sphere

500 BC: steel manufactured in India

c. 470 BC: the Ancient Greek Zeno of Elea is born

200 BC: growth of the Roman Empire

AD 876: the earliest symbol for zero is recorded in a Hindu document

c. 1,000 years ago: the tallest living tree, a *Sequoia sempervirens* that now stands at 112 metres, starts growing in California

14th century: the Black Death sweeps through Europe, killing more than 25 million people

1489: first use of + and – signs in mathematics

16th century: Leonard Digges develops the first telescopes

1543: Copernicus publishes his suggestion that the Earth goes round the Sun

17th century: Edmund Halley realises that atoms are very small, with lots of space between them

1610: Galileo looks at the Milky Way through a telescope and realises that it's made of 'innumerable' stars

1647: first map of the Moon is made by Johannes Hevalius

1680: first clocks with minute-hands appear

1687: Isaac Newton publishes *Principia Mathematica*, stating his three fundamental laws of motion

18th century: Johannes Kepler discovers the three laws of planetary motion

1703: the most destructive tsunami in history strikes Awa in Japan, killing more than 100,000 people

1705: Edmund Halley predicts the return of the comet that now bears his name

1718: Halley realises that stars move through space, and are not really 'fixed' on the sky

1775: James Watt patents his steam engine

1797: first use of iron railways, for horse-drawn wagons

1807: first use of gas to light London streets

1814: George Stephenson's first steam locomotive starts work

1859: Charles Darwin publishes *On the Origin of Species*, describing the process of evolution by natural selection

1865: automobile and motorbike are invented

1883: the volcanic island of Krakatau, near Java, erupts, killing 36,000 people in the area and more than 50,000 in the ensuing tidal wave

1889: Eiffel Tower completed

1896: discovery of radioactivity

1905: Einstein publishes his Special Theory of Relativity

1906: the most famous earthquake of all time hits San Francisco

1907: invention of colour photography

1908: a piece of a comet explodes in the air above Tunguska in Siberia, flattening 2,000 square km of forest

1912: the *Titanic* sinks

1914–18: First World War

1915: first telephone conversation between continents; Einstein presents his General Theory of Relativity

March 1918: a dangerous form of influenza originates in the USA and spreads all over the globe, killing at least 21 million people in the next four months

1920s: Harlow Shapley accurately determines the size of our Galaxy (the Milky Way)

1920s: Edwin Hubble discovers that the Universe is expanding

1923: an earthquake hits Japan, killing more than 140,000 people in fires that rage through Tokyo and Yokohama

1930: the furthest planet, Pluto, is discovered

1932: John Cockcroft and Ernest Walter use the first particle accelerator to split the atom

1939–45: Second World War

1940: first antibiotics

1943: first electronic computer

1945: a nuclear bomb destroys Hiroshima

1946: following an earthquake, oceanographers track a tsunami around Unimak Island near the coast of Alaska. This wave travels 3,620 km to the coast of Hawaii in 34 minutes – a speed of 500 km per hour

1947: first supersonic flight

1952: invention of the transistor radio; first ascent of Mount Everest

1961: first manned space flight

1965: first communications satellite successfully launched into space

2 May 1969: Neil Armstrong and Buzz Aldrin walk on the Moon; the fastest speed at which a human being has ever travelled is reached by the US crew of the command module of Apollo 10 on its trans-Earth return flight – 39,897 km per hour

1970s: increased concerns about global warming

15 April 1970: the crew of Apollo 13 fly the furthest distance from the surface of the Earth – 400,171 km out into space

1971: first pocket calculator goes on sale

1972: first e-mail software developed

1975: first personal computer available in the USA

1976: Major George Morgan and Captain Eldon Joersz of the US Air Force fly a Lockheed SR.71A aircraft at 3,529.56 km per hour; Concorde enters service

1980s: PCs become widely available

1982: the Internet is established

1996: Internet use explodes; CFCs banned throughout the developed world for contributing to ozone depletion and global warming

2002: a huge chunk of ice called the Larsen B ice shelf breaks away from the fringe of floating sea ice that surrounds the Antarctic continent

5 billion years from now: the Sun will become a Red Giant and destroy all life on Earth

Mary Gribbin works in education in East Sussex and writes books about science for children, as well as her collaborations with John Gribbin for older readers. She has long had a special interest in the history of exploration, and is a Fellow of the Royal Geographical Society.

John Gribbin is a Visiting Fellow in Astronomy at the University of Sussex, a Fellow of the Royal Society of Literature, and the author of books about science for adults. He was a member of a team that used data from the Hubble Space Telescope to determine the age of the Universe.

Ralph Edney is the brilliant illustrator of *Introducing Time*, *Introducing Fractal Geometry* and *Introducing Learning and Memory*. A cricket fanatic, he lives in London.

Nicholas Halliday lives in London and is a designer, illustrator, author and actor. His beautifully illustrated book, *21st Century Space Missions*, sold over 150,000 copies.

Index

In case of difficulty in purchasing any Wizard title through normal channels, books can be purchased through BOOKPOST.

Tel: + 44 1624 836000

Fax: + 44 1624 837033

E-mail: bookshop@enterprise.net

www.bookpost.co.uk

Please quote 'Ref: Faber' when placing your order.

If you require further assistance, please contact:
info@iconbooks.co.uk